VETERINARY ENTOMOLOGY

Livestock and Companion Animals

VETERINARY ENTOMOLOGY

Livestock and Companion Animals

Ralph E. Williams

Purdue University
West Lafayette, Indiana, U.S.A.

CRC Press
Taylor & Francis Group
Boca Raton London New York

CRC Press is an imprint of the
Taylor & Francis Group, an **informa** business

CRC Press
Taylor & Francis Group
6000 Broken Sound Parkway NW, Suite 300
Boca Raton, FL 33487-2742

First issued in paperback 2017

ISBN-13: 978-1-4200-6849-8 (hbk)
ISBN-13: 978-1-138-11818-8 (pbk)

Library of Congress Cataloging-in-Publication Data

Williams, Ralph E. (Ralph Edward), 1949-
    Veterinary entomology : livestock and companion animals / Ralph E. Williams.
        p. cm.
    Includes bibliographical references and index.
    ISBN 978-1-4200-6849-8 (hardcover : alk. paper)
    1. Veterinary entomology. 2. Insect pests--Control. 3. Livestock--Parasites. 4. Domestic animals--Parasites. I. Title.

SF810.A3W556 2009
636.089'6968--dc22   2009039059

Visit the Taylor & Francis Web site at
http://www.taylorandfrancis.com

and the CRC Press Web site at
http://www.crcpress.com

DEDICATION

I dedicate this book to my wife Jeanine and children Jason, Sarah, Amy, Nicholas, and Clayton, and to the loving memory of my son John. I also dedicate this book in the memory of Dr. R. C. Dobson who inspired me to pursue a career in veterinary entomology, and to Dr. E. C. Turner, Jr. and Dr. Jakie A. Hair, who provided me the opportunity to fulfill my goal in this field. In addition, I dedicate this book to Dr. Fred Knapp and Dr. Jack Campbell, who have been role models, mentors, and colleagues from whom I have learned much throughout my career in veterinary entomology.

CONTENTS

Contents

LIST OF ILLUSTRATIONS

PREFACE

Veterinary Entomology: Livestock and Companion Animals is a textbook/reference written for undergraduate students interested in this field. Its intent is for use as a suitable textbook for courses dealing with veterinary entomology. It is designed to acquaint the student with the array of insects and related arthropods that may be encountered in animal husbandry practices and with companion animals. The text provides pertinent information on host–pest relationships, pest control methodology, identification and biology of important pest groups, and the importance and impact of pests within each livestock commodity and companion animal group.

Not included in this textbook is any detailed information on structure and function of insects and related arthropods or a review of classification of arthropods. This information is readily available in numerous other textbooks on entomology and is not directly pertinent to the subject matter being presented in this textbook. Information regarding identification, ecology, lifecycles, and so forth, is presented, as necessary, in chapters dealing with specific pest groups.

This textbook is suitable not only for instruction of courses at the college level, but also as a ready reference for anyone needing information on arthropods affecting livestock and companion animals. Veterinarians, animal scientists, extension agents, industry technical representatives, and livestock producers, among others, should find this book useful for obtaining information on pests of concern.

Updated trends in the field of veterinary entomology are presented in this text. Origins of an earlier textbook (Williams et al. 1985) came from a group of participants who regularly attend the Annual Livestock Insect Workers Conference, an informal gathering of university researchers, extension specialists, students, government personnel, and industry personal. This conference has been held yearly for over 50 years at various locations around the United States. Many changes have been made in veterinary pest management since

1985 when this first book was published. Those new trends in pest management practices and new pest problems are addressed in this textbook.

Because of frequent changes in specific chemicals and pest procedures, no discreet chemical control recommendations are provided in the textbook, other than where example methods may be discussed. Control recommendations are routinely revised by federal and state agencies because of product and/or methodology changes that frequently take place. Individuals needing current information on effective and safe pesticides and control recommendations are encouraged to call, write, email, or visit Web sites of their local extension agent or state extension entomology specialist.

All attempts have been made to ensure that the information presented in this textbook is as accurate and reliable as possible. I encourage anyone who finds any errors or misstatements to bring these to my attention.

Ralph E. Williams, PhD, D-ABFE

ACNOWLEDGMENTS

The author wishes to thank all of those dedicated individuals who have committed themselves to the field of veterinary entomology, including those in education and research, government, and industry. A special thanks goes to those individuals who continue to support and attend the annual Livestock Insect Workers Conference (LIWC) to keep us abreast of current trends in this field.

Special thanks go to Tammy Luck, assisted by Daniel Dugard and Sarah Cox, for converting artwork and slides to digital format for inclusion in this book. Special thanks also go to Jim Kalisch, University of Nebraska, for spending the time to research and provide several of the photographs used in this book. The author is also grateful to the following individuals and organizations providing illustrations and photographs reproduced in this book: Joseph Berger (Bugwood.org), John B. Campbell (University of Nebraska), Scott Charlesworth (Purdue University), Commonwealth Scientific and Industrial Research Organization (CSIRO) (Australia), Marcelo de Campos Pereira (University of Sao Paulo, Brazil), Michael Dryden (Kansas State University), James Gathany (Centers for Disease Control and Prevention [CDC]), Peggy Greb (USDA-ARS), Robert Hall (University of Missouri), Patrick Jones (Purdue University), John Kucharski (U.S. Department of Agriculture, Agricultural Research Service [USDA-ARS]), Lincoln University (Natural Science Image Library), Virginia Martin (www.feathersite.com), Novartis, Richard Nowitz (USDA-ARS), John Obermeyer (Purdue University), Oklahoma State University, Univar, University of California, Steve Upton (Kansas State University), and USDA-ARS.

The author would also like to thank the staff at Taylor & Francis Group, especially John Sulzycki and David Fausel.

INTRODUCTION

Veterinary entomology encompasses the study of insects and related arthropods that have a direct effect on the health and well-being of domestic animals, namely livestock raised for food production and companion animals (horses and pets). It is a branch of the broad scope of veterinary parasitology specifically focused on arthropods. The related field of medical entomology focuses on those arthropods of importance to human health. In some instances, the same arthropod groups and species can affect both humans and animals. And, livestock production systems can produce medically important entomological concerns (e.g., mosquito breeding sites and nuisance fly concerns). Also, some arthropod-borne diseases such as West Nile virus and some types of encephalitis can affect both humans and animals. Emphasis in this textbook is on those arthropods specifically associated with livestock and poultry production systems and those arthropods affecting companion animals. In situations in which livestock husbandry practices may contribute to medically important entomological problems, this is discussed in the appropriate chapters.

An integral part of comprehensive animal-health programs and production management practices is an understanding of veterinary entomology. Insects and related arthropods can affect animals directly by feeding upon or invasion of animal tissue, exsanguination from blood sucking, and introduction of venoms or salivary secretions. Indirectly, arthropods can serve as vectors of pathogenic organisms. In addition, many species of arthropods that originate from various animal husbandry production practices may constitute annoyance to farm workers, nearby residents, or livestock or companion animals directly.

Veterinary entomologists and those who professionally deal with arthropods associated with livestock and companion animals can be found in an array of jobs within the U.S. Department of Agriculture (USDA), state agricultural experiment stations, State and Federal Cooperative Extension Service, colleges and universities, private industry, and independent private consulting and contracting companies.

Veterinary entomologists address all aspects of arthropods of veterinary importance from identification, biology, distribution, abundance, economic impact, pest management practices, and pest control product development. In this diverse field, many of the personnel involved may specialize in a particular group of arthropods, specific livestock or companion animal groups, production management systems, or specific control products or methodology.

HISTORY

There are many historical and ancient references of arthropod pests of domestic animals. However, it wasn't until the late 1800s that the modern genesis of veterinary entomology gained recognition. In the 1890s, USDA researchers Theobald Smith and F. L. Kilborne established that the pathogen causing Texas cattle fever was transmitted by the tick species *Boophilus annulatus*, subsequently called the Texas cattle fever tick. This was the first documentation of the transmission of a disease agent by an arthropod causing a disease in livestock. By the turn of the 20th century, knowledge relative to veterinary and medical entomology advanced rapidly. Several notable research efforts paved the way for the modern era of veterinary and medical entomology. Some notable examples include the pioneering work by Ronald Ross, who, in 1898, working in India, demonstrated the role of mosquitoes in transmission of the causative agent of malaria; studies initiated by Carlos Finlay, followed by Walter Reed in 1890 with the U.S. Yellow Fever Commission, documenting the association of mosquitoes in the transmission of yellow fever; discovery by David Brue in 1903 that the trypanosomes that caused nagana (in Africa) were vectored by *Glossina* sp. flies; and Walter Reed and others on the association between house flies and such disease agents as typhoid bacilli and enteric diseases. At about the same time frame, entomologists including such individuals as Charles V. Riley, Leland O. Howard, F. C. Bishop, and E. W. Laake were instrumental in building the groundwork of modern veterinary entomology going into the 20th century. In 1896, a notable book was published by the USDA's Division of Entomology by Herbert Osborn titled *Insects Affecting Domestic Animals*. With this publication and his early contributions to the field, Osborn is considered as the "Father of Livestock Entomology."

In the early 20th century, there were increased hirings of veterinary entomologists by both federal and state agencies. Developments were made in the chemical industry with new pesticides being developed (i.e., botanicals, chlorinated hydrocarbons, carbamates, organophosphates) which created new approaches to veterinary pest control. By 1946, the USDA established the U.S. Livestock Insect Laboratory in Kerrville, Texas. At this laboratory, in the early 1950s, R. C. Bushland, in conjunction with E. F. Knipling, demonstrated that viable sterile male screwworms, *Cochliomyia hominivorax,* could be produced. This knowledge eventually led to the eradication of the screwworm from the United States, Mexico, and Central America.

CURRENT STRUCTURE/OPPORTUNITIES OF VETERINARY ENTOMOLOGY IN THE UNITED STATES

Federal Organizations

Now known as the Knipling-Bushland U.S. Livestock Insects Research Laboratory, the focus of the Kerrville laboratory today and its off-site locations is to provide the U.S. cattle industry and the public technology for eradicating or controlling pests (primarily ticks, blood-feeding flies, and screwworms) of veterinary and medical importance. The USDA also has additional locations where veterinary entomology research is being conducted. Scientists at the USDA Livestock Insects Research Unit at the University of Nebraska–Lincoln conduct research on the stable fly, *Stomoxys calcitrans*, a pest species found in abundance near the cattle feedlots on the Great Plains. At another USDA laboratory, Center for Medical, Agricultural, and Veterinary Entomology, in Gainesville, Florida, research, in part, is done with arthropods of veterinary importance.

State and Other Organizations

Many states in the United States employ entomologists with at least some responsibility in veterinary entomology. Most are at land-grant universities that foster agricultural research, extension, and teaching. Medical and veterinary entomology courses are taught at a majority of these universities. Several courses dealing specifically with livestock pests are also being taught. Most of the teachers involved with these courses, along with others who do not have teaching appointments, conduct research on problems in veterinary entomology pertinent to their state or region. In addition, most land-grant universities have cooperative extension service personnel responsible for veterinary entomology pest problems within their state. Extension entomologists assess and transfer research results and technology to other state specialists, local agricultural agents, and the general public. Entomologists and others with expertise in veterinary entomology may also be located at non-land-grant state universities, private colleges, and various levels of state government.

Within private industry, there are many veterinary entomologists employed by companies engaged in the development of pest control products and methodologies, in research and development, technical support, or marketing. With the trend of many of the older products being eliminated from the marketplace because of environmental and health and pesticide resistance concerns, many new products and methodologies are emerging and continually being developed by private industry.

With recent cutbacks with some universities and state and federal agencies, there are more veterinary entomologists getting involved in private consulting practices and independent research. Trained professionals with knowledge of pest species and various livestock, poultry, and companion animal production and management systems are providing clientele needed information on veterinary pest management issues. Also, with the trend of urban sprawl and more nonfarming residences moving into agricultural districts, there has been a need for trained veterinary entomologists to work with livestock producers to minimize nuisance and health risks associated with insects (i.e., flies, litter beetles). In some cases, veterinary entomologists have been called upon as expert witnesses in legal matters dealing with nuisance pest issues.

EDUCATION NEEDS/OPPORTUNITIES

The training and knowledge needed for competency in veterinary entomology involve both undergraduate- and graduate-level education. Many universities offer undergraduate programs in entomology. Other disciplines of study in zoology, biology, animal science, and veterinary science offer students an undergraduate education that can provide the student basic learning skills that could lead to a career in veterinary entomology. At the undergraduate level, a well-rounded academic plan of study should include the basic sciences (biology, zoology, chemistry, physics), mathematics, and communication (oral, written) skills. In addition, depending on the student's interests, course work in other disciplines is recommended. These often include courses in invertebrate zoology, ecology, virology, bacteriology, parasitology, microbiology, insecticide science, animal production courses, statistics, and climatology, among others. Most colleges and universities have a suitable array of courses that can provide basic education in most of these areas.

Graduate-level education is generally required for one to be recognized as a veterinary entomologist. Many universities offer graduate programs in entomology that have faculty trained as veterinary entomologists. It is generally recommended that the student seek both the master of science (MS) degree and the doctor of philosophy degree (PhD). These degrees should be thesis and dissertation based. Master degree programs normally require 2 years of training with course work and research project. Course work is often tailored to the student's need to gain proper working skills and knowledge not obtained at the undergraduate level and that may be needed for the designated research project. Advanced courses in entomology, parasitology, statistics and experimental design, and animal product systems, among others, are courses that are often part of a MS degree program. In the PhD program, the student hones in on skills to be recognized as a veterinary entomologist. The PhD degree typically requires an additional 2 to 3 years of course work and research beyond the MS degree. If the student pursues the PhD directly after the undergraduate degree, it can take 3 to 5 years to obtain. The plan of study for the PhD student is usually tailored to the student's needs and career interests, along with additional needs for the research area studied. The research at the PhD level involves a substantial piece of original scholarly research presented in the form of a written dissertation. This research should be of scientific sound quality to be published in a peer-reviewed scientific publication. The passing of oral and written examinations in addition to satisfactory completion of course work and thesis or dissertation is generally required for the successful completion of a postgraduate degree program.

AVAILABLE SOURCES OF INFORMATION

Several sources of information are available for obtaining information on research findings and pest management information on insects and related arthropods of veterinary importance. Original research results are generally published in peer-reviewed scientific journals. Journals that frequently have veterinary entomology research findings include the *Journal of Economic Entomology, Annals of the Entomological Society of America, Environmental Entomology, Journal of Medical Entomology, Journal of the Kansas Entomological Society, Southwestern Entomologist,* and *Florida Entomologist,* among other entomology-specific journals. Scientific papers dealing with veterinary entomology are also frequently published in other journals, including *Veterinary Parasitology, Journal of Parasitology, Parasitology, Experimental Parasitology,* and in various professional veterinary journals.

Current trends, news, and control recommendations of insects and related arthropods of veterinary importance can be obtained from USDA bulletins, state extension publications, and trade journals of the different livestock, poultry, and companion animal industries. There are also numerous Web sites from federal and state institutions and private industry providing a wealth of information on veterinary insects and related arthropod pests.

THE AUTHOR

Ralph E. Williams, PhD, D-ABFE is a professor of entomology at Purdue University, West Lafayette, Indiana, and the Director of the Purdue Forensic Science Program. His area of focus is with arthropods of public health and veterinary importance and forensic entomology. As a veterinary entomologist, his emphasis is placed on addressing the continuous need for development of pest management strategies relative to the protection and productivity of livestock and poultry. Forensic entomology research focuses on the role of forensically important insects in death investigation. Williams also coordinates the teaching of the minor in forensic science at Purdue. He has a BS degree from Purdue University, an MS degree from Virginia Tech University, and a PhD degree from Oklahoma State University. He is board certified by, and a diplomate of, the American Board of Forensic Entomologists. Williams holds membership in several scientific associations and has authored more than 35 scientific publications, authored or coauthored four books, and has made more than 100 presentations at scientific meetings.

CHAPTER 1

IMPORTANCE OF ARTHROPODS

TABLE OF CONTENTS

A diverse array of arthropod species has developed relationships with other animals in one way or another. Direct relationships in which an organism (known as a parasite) derives some benefit from such a relationship to the detriment of the organism (known as a host) is called *parasitism*. The host provides several resources for parasites. The host may supply a source for food (i.e., blood, lymph, body secretions, skin debris, hair, feathers). The host's body can also provide an environment for arthropods to live in providing warmth, moisture, and protection.

Arthropods parasitize a wide range of hosts, including other arthropods. Those arthropods that spend all or some portion of their lives affecting the well-being of livestock, poultry, or companion animals are discussed in this book. They include mites, ticks, lice, fleas, true bugs, beetles, and flies. Other groups of arthropods considered as beneficial in controlling pest arthropods are also discussed.

HOST–PEST ECOLOGICAL RELATIONSHIPS

The two classes of organisms in the phylum Arthropoda of importance to livestock, poultry, and companion animals include Arachnida (scorpions, mites, ticks, spiders) and Insecta (insects). Insects make up 70% to 75% of all known animal species.

Within the class Arachnida, mites and ticks are of the most veterinary importance. Most mite species are free living, but some are parasitic on wild and domestic animals, such as mange mites on livestock and mites that attack poultry. All tick species are parasites of vertebrate animals.

Within the class Insecta, there are four orders of insects that are of primary importance as pests of domestic animals. These include Phthiraptera (Anoplura—sucking lice, and Mallophaga—chewing lice), Siphonaptera (fleas), Hemiptera (true bugs), and Diptera (true flies). All species of Phthiraptera (including Anoplura and Mallophaga) and Siponaptera are parasites of vertebrate animals. There are several species of both sucking lice and chewing lice parasitic to domestic animals. Most lice species are considered host specific, primarily found on only one host species or closely related species. Within Siphonaptera, there are several flea species that attack domestic animals. However, only a few species are of significant veterinary importance. Within Hemiptera are a few species (e.g., bed bugs) that are parasitic on vertebrate animals. The order Diptera contains the most numerous and diverse species of insects that have economic importance as veterinary pests. Some Diptera suck blood as adults and include important pests of both humans and livestock (e.g., mosquitoes, black flies, horse and deer flies, stable flies, horn flies, sheep keds). Other Diptera do harm in their larval stages by invading flesh and tissue of animals (e.g., cattle grubs, horse bots, sheep nose bot, screwworms). Other Diptera contribute to nuisance and annoyance to livestock and humans (e.g., house flies). In addition to direct parasitism, many of these arthropod groups also serve as vectors of disease agents, and some may be considered of importance just by their presence.

Types of Arthropod Parasitism

Parasitic arthropods can be classified in various ways by body site occupied, their relationship and dependency with a host during all or part of their lifecycle, and host specificity.

Some arthropod parasites of domestic animals are considered as internal parasites, spending at least part of their life in body tissue of the host. Of most importance to domestic animals are the myiasis-producing flies. These include flies in the families Calliphoridae (blow flies, screwworms), Sarcophagidae (flesh flies), Oestridae (cattle grubs, sheep nose bots), Gasterophilidae (horse bots), and Cuterebridae (rodent bots).

External parasites (also called ectoparasites) obtain nourishment from their hosts on the skin surface, either from sucking blood and body fluids or feeding on skin debris, hair, or body secretions. Some arthropods such as sarcoptic mange mites burrow under the skin just under the surface and are also considered external parasites.

When a parasite is dependent on a host for some resource for continued life or to complete a portion of their lifecycle, this is referred to as obligatory parasitism. Facultative parasites, on the other hand, may feed or live only occasionally on a host and are not dependent on the host for survival.

With obligatory parasitism, there can be considerable variation in the amount of time spent in or on the host. Some parasites live in continuous association with their host throughout their entire lifecycle. They are highly dependent on the host for survival. Lice, some species of mites, and sheep keds are examples of continuous parasites. They are usually disseminated between individual hosts by direct contact. Most parasitic arthropods have only intermittent contact with their host and are free living for a major portion of their lifecycle. Only certain life stages of these parasites depend on the host for resources. An example of this intermittent type of parasitism is hard ticks (family Ixodidae). Eggs are most frequently laid on the ground by the adult female tick. Upon hatching, the larval ticks seek a suitable host to feed upon. Some tick species will then remain on the one host to complete their nymphal and adult molts (one-host ticks); in other species, each stage of tick drops from one host to molt into the next stage before seeking another host. Among these are three-host ticks in which larvae, nymphs, and adults feed on different individual hosts. Also, in most blood-sucking flies, only the adults are parasitic, with the other life stages being free living.

Host specificity varies among parasitic arthropods. With some arthropods, only one host species is exploited. Many species of lice, mites, and some Diptera, such as keds and cattle grubs, are typically found on a particular type of host animal. As an example, the hog louse will infest swine and no other animals. The two species of cattle grubs found in the United States will only successfully develop in cattle, although occasional infestations may be found on horses and other animals. In these cases, the cattle grubs usually do not reach maturity in the nonbovine host. Horse bots are less host specific and may successfully complete their lifecycle on several species of equines. Ticks, especially three-host ticks, will usually feed on a multitude of different host species, depending on their life stage. Larvae and nymphs of some species may feed on various ground-nesting birds and rodents, whereas adults may feed primarily on larger hosts such as cattle. Several of the blood-feeding flies, such as mosquitoes, tabanids (horse and deer flies), and stable flies, will feed on several host species, whereas horn flies prefer cattle as their primary blood-feeding resource.

The type of association a parasitic arthropod has with its hosts, whether it be considered internal/external, obligatory/facultative, or continuous/intermittent, and to their degree of host specificity can have important implications to pest control programs and in the treatment of arthropod-borne diseases. For example, controlling hog lice on swine is often a much easier task because of their being continuous external parasites and host specific, as compared to controlling certain ticks, such as lone star ticks, that have free-living stages and may have multiple domestic animal hosts and wild hosts.

Many disease-producing organisms not only can be transmitted from one host domestic animal to another, but may be reservoired in wildlife populations. Arthropods that may associate between certain wildlife species and domestic animals can be responsible for the movement of these disease agents between these groups of animals and even humans. Examples of such diseases and arthropod vectors include encephalitis and mosquitoes, and Lyme disease and ticks.

Environmental factors will have influence on the various types of parasitism. Continuous parasites that have a direct contact with the host are afforded an abundant food source, warmth and moisture, and optimal mating opportunity. Intermittent parasites, on the other hand, are subject to environmental stresses, especially in their free-living stages. Finding suitable mates and breeding habitats and nonhost food sources often influence parasite propagation and survival.

Availability of suitable breeding habitats is a very important limiting factor for many parasitic species in the free-living stages. Aquatic habitats such as waste lagoons, streams, swamps, marshes, wetlands, and numerous other aquatic habitats are needed for larval development of many intermittent parasites, especially for such biting fly groups as mosquitoes, tabanids, biting midges, and black flies. Suitable supplies of organic material and animal wastes are needed for other pest species, such as stable flies and house flies. An important facet of pest management is alteration, or elimination, of these various breeding habitats, making them unsuitable or unavailable for pest development, and helping break the pest's lifecycle. A major component in confined livestock and poultry operations is manure and waste management as an integral part of the pest management program.

PROBLEMS CAUSED BY ARTHROPODS

Arthropods associated with livestock, poultry, and companion animals cause problems in a variety of direct and indirect ways. Direct damage by parasitic arthropods ranges from bite irritation, blood loss, tissue consumption and damage, and toxic and allergic reaction to arthropod venoms and feeding antigens. Indirect damage may be caused by introduction of pathogens, nuisance leading to stress and annoyance, self-inflicted wounds from scratching and pest avoidance, and in creating susceptibility to other stressors. Peripheral effects such as filth fly nuisance and affect of pesticide use and exposure on the general health of animals may also be of concern.

Direct Damage

Blood loss caused by blood-feeding arthropods can be of economic concern to animal productivity and well-being. Although an individual blood-feeding arthropod may consume only a small amount of blood from its host, in large numbers they may contribute to significant blood loss. Certain tick species, as an example, such as the lone star tick, *Amblyomma americanum*, when found by the thousands on a single host, can cause considerable blood loss and anemia to their host. It has been shown that over 90 kg of blood can be removed by ticks on cattle over a single season. Tabanid horse flies can consume up to 0.5 L of blood a day from their host. Large flea infestations can result in significant blood loss (~10%) in cats and dogs over several days. Associated with blood-feeding activity can be toxic and allergic reactions caused by antigens and anticoagulants in salivary secretions. Toxins in the saliva of certain ticks, as an example, may produce a host paralysis that can be lethal if not treated.

Bite irritation causing skin inflammation and pruritus (itching) can be caused by arthropod feeding activity. This can lead to hair/wool loss, skin thickening, and secondary infections.

Myiasis, the infestation of tissue, can be caused by various arthropods, including certain mites, biting lice, tissue-invading flies, and other parasitic arthropods that consume or harm tissues of their hosts. Skin damage and hair loss often result from sarcoptic and psorpotic mite infestation. Cattle grubs, *Hypoderma* sp., cause considerable damage to tissue and skin of cattle during their parasitic migration through their host. Primary screwworms, *Cochliomyia hominvorax*, consume living tissue, resulting from fly eggs being laid on a wound or other break in the skin. Secondary screwworms, *C. macellaria*, and other blow flies that feed on dead tissue can cause host trauma and secondary infection to their host. Lice infestation can cause thickened skin and hair loss.

Indirect Damage

One of the most important issues associated with arthropods is the ability of some to transmit disease-producing pathogens. Such arthropods are known as vectors. Pathogens transmitted by arthropods may include protozoa, bacteria, viruses, tapeworms, and nematodes. Arthropods may transmit these pathogens by biological or mechanical means.

In biological transmission, there is biological development of the pathogen inside of the arthropod after the arthropod acquires the pathogen from an infected animal. After appropriate development of the pathogen, the vector becomes infective and can transmit the pathogen to another animal the next time it feeds. In this situation, the arthropod serves as an intermediate host and is known as a biological vector.

In mechanical transmission, the vector serves only as a mechanical carrier of the pathogen—the pathogen is usually adhering to the vector's mouthparts, body, or feet. The vector picks up the pathogen while feeding on an infected host and carries the pathogen to another host. Successful transmission of the pathogen to the second, uninfected host usually occurs within a few hours, because longevity of most pathogens is relatively limited when exposed outside their hosts.

In livestock and poultry production and among companion animals, arthropod-borne diseases can be of major concern. In some extreme situations, such as in Africa, tsetse flies (*Glossina* sp.) serve as vectors of trypanosomes that cause an illness known as nagana in domestic animals. Nagana has prevented extensive cattle production in much of Africa. Historically, in North America, Texas cattle fever, transmitted by the cattle fever tick, *Rhipicelphelus (Boophilus) annulatus*, impaired much of the cattle industry in the Southern United States and Mexico in the early 1900s. Eradication efforts were successful in eliminating this tick vector from most of its northern range, which led to an effective control of disease occurrence.

There are several arthropod-borne diseases of concern today in domestic animals. Viral mosquito-borne encephalitides can affect both humans and domestic animals, especially horses. A recent mosquito-borne disease threat is that of West Nile virus. This disease has had severe consequences to both humans and horses throughout the United States. Viruses causing bluetongue in sheep and epizootic hemorrhagic disease (EHD) in cattle are transmitted by biting midges (*Culicodes* sp.). Among other diseases, the protozoan that causes anaplasmosis in ruminants is mechanically transmitted by some species of horse flies and deer flies. The bacterium that causes infectious bovine keratoconjunctivitis (IBK) (or "pinkeye") in cattle is mechanically transmitted by adult face flies as they feed on eye secretions of cattle. Among companion animals, the nematode that causes dog heartworm is biologically transmitted by various mosquitoes. Further information on specific arthropod-borne diseases is presented in chapters dealing with each arthropod group.

Livestock, in attempts to avoid or escape from arthropod attack, can be injured or be less productive. The activity of certain arthropods, such as adult warble flies (*Hypoderma* sp.), often leads to dramatic avoidance responses. *Gadding* is the term used to describe the panicky avoidance behavior in cattle being attacked by warble flies. Cattle run madly around from the buzzing sound of these flies with their tails straight up. Self-injury often occurs following collisions with fences and other objects. In other situations, cattle and other livestock will bunch together, go into ponds, lay in ditches, and such to avoid fly annoyance. Also, it has been shown that "fly worry" can lead to less grazing time and less feed efficiency. Decreased vigor and increased susceptibility to environmental stress and other diseases often result. Skin damage, either by direct feeding of arthropods or injury from avoidance, can also allow for secondary infections caused by bacteria and other pathogens.

The consequences to livestock, poultry, and companion animals to arthropod attack and annoyance leads to reduced productivity, animal well-being, and profitability. Reduced weight gains and feed efficiency result from animals needing more energy to replenish lost blood, repair damaged tissue, and engage in avoidance behavior that limits grazing activity. Many studies have shown how arthropods can have direct effects on animal productivity: stable fly activity resulting in reduced weight gains and feed efficiency in beef cattle and less milk production in dairy cattle; ticks and horn flies affecting cattle weight gains of pastured cattle; hog lice and sarcoptic mange mite infestations affecting swine productivity and reproductive capacity; northern fowl mite infestations reducing egg production in chickens; and sheep keds reducing wool production, among other studies. Animals affected by arthropod attack often show signs of less vigor and body condition, which can reduce their desirability and profitable sale values.

Other indirect consequences of arthropods associated with livestock and poultry production are environmental nuisance and product downgrading. Large populations of flies and litter beetles that may breed in animal wastes may cause nuisance concerns on properties surrounding the production site. Also, as animal wastes are removed and spread on agricultural land, these pests often leave the field to invade nearby properties. Nuisance complaints and health concerns have often led to legal complaints and civil lawsuits. In one case in a Midwestern state, a large egg producer lost a civil lawsuit amounting to over $20 million due to fly nuisance. With the trend of more confined livestock feeding operations and large-scale dairy and poultry production systems being established, and the rapidly growing urban and suburban sprawl into agricultural districts, the concern of insect nuisance, odors, and groundwater contamination will be a continued concern. As to product downgrading, in dairy and egg production, for example, dairy and poultry plant inspectors may downgrade milk and eggs to lesser grades if too many flies are present in milking parlors or processing areas. It is a wise management decision to incorporate a well-defined pest control component into the standard operating procedures of a livestock or poultry production operation to avoid costly legal issues and to strive for more efficient and profitable productivity.

The effect that arthropods have on domestic animals can often be assessed as to economic injury levels and threshold levels. The economic injury level (EIL) represents the point at which the damage or reduction in yield equals the cost of control. It is the smallest population of the pest which will cause economic loss to the producer. It is at this point that the producer should strive to prevent pest populations from rising above this level. As illustrated in Figure 1.1, the economic threshold level (ETL) is the population density of the pest just below the EIL at which measures should be taken to prevent an increasing pest population from reaching the EIL.

Although ETLs have been used extensively in agronomic cropping systems for pest management practices, it is much more difficult to obtain accurate values for animal production systems. Animals are quite complex in their responses to arthropod attack. In theory, the larger the pest population, the greater is the potential for reduction in animal productivity. Most animal species can withstand small numbers of ectoparasites without showing signs of adverse affects. Likewise, very large populations of ectoparasites often produce

7

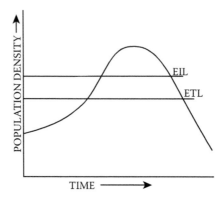

Figure 1.1
Economic injury level (EIL) and economic threshold level (ETL) over time and pest population density.

proportionally less production loss per parasite than smaller numbers. Reasons for this often relate to the host's physiological reaction to parasitism levels, with higher numbers of parasites not being able to obtain optimal nourishment from their host. Also, environmental conditions may limit the ability of parasites to effectively feed on the host at higher levels. Cooler temperatures may serve in reducing parasite activity, as an example. The impact of ectoparasites on animal performance is best envisioned by an "S-shaped" curve in which there are limited effects of small parasite populations, and a leveling off of effects with larger populations. The key is to determine the range in between where most harm to the host is done. This is where the determination needs to be made of the ETLs.

Various factors need to be taken into account for determining the ETL of a particular pest or pest group. Host age may be of influence. Younger hosts often are more susceptible to parasite attack than older hosts. Older animals often build up a degree of natural immunity to parasite attack. Host nutrition can affect the success of parasites to do harm to their host. Animals kept on higher planes of nutrition can often tolerate larger parasite loads or reduce host physiological reactions to parasite attack. As an example, forage quality can influence cattle nutrition and ability to withstand parasite attack. Related to forage quality is the nature of nutrients present in animal manure that may impact manure-breeding parasites. Horn flies and face flies prefer manure from grass-fed cattle to cattle that are on a higher plane of nutrition such as grain-fed cattle. Soil type and environmental conditions (temperature, precipitation, humidity, etc.) can also contribute to forage quality and animal nutrition.

Host resistance to parasites raises the ETL. Not only do some susceptible hosts have the ability to develop natural immunity, but certain breeds of livestock are able to withstand ectoparasite attack better than other breeds. Brahman cattle, *Bos indicus*, have historically shown to be able to resist the damaging effects of ticks and other ectoparasites better than European

Figure 1.2
Bos indicus *and* B. taurus *cattle tick breed resistance research.*

breeds, *Bos taurus*. *Bos indicus* cattle and crossbreeds with *Bos indicus* breeding have often been the preferred breeds in warmer climates and in areas where ectoparasite pressure is higher (Figure 1.2). In warmer, southern states where ticks are prevalent on cattle, *Bos indicus* and crossbreeds have been shown to tolerate tick attack better than European breeds. Similar trends have been accomplished in using strains of chickens resistant to northern fowl mites. Development of genetic-resistant strains of livestock and poultry is a promising alternative in integrated pest management programs.

Other complications in determining the ETL come into play with the presence of multiple ectoparasite attack. Often, more than one species of lice may be found feeding on cattle, swine may have both lice and sarcoptic mange mites present, sheep may have simultaneous sheep ked and lice infestations, and so forth. In addition, seasonal trends in parasite abundance will influence ETL levels. Lice on cattle, as an example, tend to be more prevalent in colder months, whereas horn fly populations are more prevalent in warmer temperatures. Different geographic areas may also vary in influencing which ectoparasite species are more prevalent. For instance, face flies on cattle are more prevalent in temperate regions of the United States and are almost nonexistent in warmer, southern states.

In addition to the effects that arthropod attacks have on hosts, there are other economics involved in the costs of control efforts. There are costs associated with pesticides and other control products, costs for application equipment and maintenance, costs for suitable livestock handling facilities and maintenance, and cost of labor to carry out control programs. There is often a need for subsequent or multiple treatment applications. Reduced animal performance can result due to pesticide application trauma and handling stress. Arthropod resistance to pesticide usage can occur, such as with horn flies on cattle. With the use of pyrethroids in insecticide-impregnated ear tag devices, widespread horn fly resistance to the pyrethroids being used has developed to the extent that such treatments have become ineffective in many areas of the United States. Additional costs are passed on to the animal producer or owner with the costs of development of new pesticide products and control strategies. Additional cost factors may arise from possible pesticide residues in meat and other livestock and poultry products.

CHAPTER 2

PRINCIPLES OF ARTHROPOD MANAGEMENT

TABLE OF CONTENTS

The primary reasons for pest control activities in veterinary entomology are to reduce or prevent losses of food and fiber; maintain healthy livestock, poultry, and companion animals; and reduce insects associated with animal production practices as a source for public nuisance.

In carrying out any successful control program, certain principles need to be followed:

1. Correct identification should be made of the pest species involved.
2. A basic understanding of the biology, lifecycle, and habits of the pest is needed in order to assess at which vulnerable stage or stages of the pest direct control efforts should be applied.
3. An assessment should be made of the magnitude of damage done by the pest. This is determining the economic threshold level (ETL) of the pest for deciding when control action should be taken.
4. Estimates should be made as to the amount that can be spent profitably in reducing the damage caused by the pest.
5. Determination then needs to be made on the execution of the most efficient approach to controlling the pest. Effective control must be economically sound for both immediate results and long-term effects.

Historically in veterinary entomology there was a strong tendency toward dependence on chemical pesticides, especially in the early and mid-1900s. With concerns emerging in the 1960s and 1970s of environmental consequences of indiscriminate pesticide use and with the subsequent establishment of the U.S. Environmental Protection Agency (EPA), changes started taking place in pest control practices. Emphasis was being placed on integrated pest management systems to strive for reductions in pesticide usage and optimization of pest control efforts. Several problems have been identified with pesticide dependence in pest control. There is concern of environmental buildup of pesticide residue. There exists the possibility of food contamination. Pest resistance to insecticides is of concern. Cancellation of approved pesticides has been a deterrent, especially in situations where few pesticide registrations exist for particular pest groups. Also, increased costs of control have been incurred with development and registration costs in bringing new products on the market.

The goal of a sound pest control program in veterinary entomology is to strive for an integrated approach to reduce pest populations below the ETL. In doing so, various control components (biological, cultural, chemical, mechanical) are utilized in harmony to maximize profitability and minimize environmental impact. This is considered the primary goal of a sound integrated pest management program. It is the selection of and integration of arthropod control methods on the basis of anticipated economic, ecological, and sociological consequences.

Consequences of sound integrated pest management programs in animal production include several benefits. Safer and more effective pesticides and control techniques are utilized. There is resultant less pollution caused by indiscriminate pesticide usage. More, cheaper, higher quality, and efficiently produced meat, poultry, and dairy products are made available for the consumer with less potential chemical contamination. A reduction in veterinary services and expenses is realized. And, overall, there are healthier livestock, poultry, and companion animals.

There are various tools and methods that make up integrated pest management programs for animals. These start with proper surveillance techniques for assessing pest population levels present and damage done. Then the various methods of carrying out pest management include cultural, mechanical, physical, biological, chemical, genetic, and regulatory methods.

PEST SURVEILLANCE

An initial effort in pest management is to establish the extent of pest infestation and damage levels in animals. Surveillance efforts are made to identify species of arthropods present, assess their distribution, establish host associations, and determine pest population densities at specific times and locations. Initial surveys should be done to establish the current situation followed by an ongoing surveillance activity to assess pest population fluctuations over time and to assess pest reduction efforts resulting from pest management activities. This will allow for any necessary adjustments that may be required to optimize control strategies. Surveillance will range from arthropod trapping methods, use of insect nets, direct animal examinations, and even observation of animal behavior in reaction to arthropod attack. Surveillance techniques need to be established specific to the pest group and animal production system. Surveillance procedures for specific pest groups are discussed in later chapters in this book.

CULTURAL METHODS

Proper cultural practices in animal production can serve not only to reduce actual pest populations but to minimize situations where pest species can reproduce. With confined livestock and poultry production, waste management practices can effectively reduce fly breeding and fly activity. Such fly species as house flies (*Musca domestica*) and stable flies (*Stomoxys calcitrans*) are of major concern around such production systems. House flies

Figure 2.1
Removing manure in an egglayer house pit.

prefer moist fecal material, whereas stable flies will breed in moist animal manure mixed with soil and decaying feed and moist fermenting plant material. Proper waste management practices of regular waste removal and keeping moisture levels down can prevent major concerns for nuisance fly populations (Figure 2.1). Various activities to curb moisture levels in confined livestock and poultry operations range from preventing water leaks, designing proper floor grade to allow adequate moisture drainage, providing adequate ventilation for waste drying and animal comfort, avoiding high temperatures to reduce animal need for excessive water intake, and using absorbent litter where feasible.

Another cultural practice to reduce parasitism is to timely rotate pasture-grazed livestock from one pasture to another (Figure 2.2). This aids in reducing the incidence of some internal parasites and ectoparasites such as ticks. Other practices that can help reduce parasitism include proper timing of calving to minimize attack by flies and proper timing of such practices of dehorning, castration, and wool shearing to prevent fly attack of wounds. Proper outdoor water management is yet another practice to reduce mosquito breeding and the incidence of horse flies, deer flies, black flies, and other blood-feeding aquatic insects prevalent during warmer weather months.

Figure 2.2
Cow pasture.

MECHANICAL AND PHYSICAL METHODS

Proper feedlot design and maintenance are very important to prevent and minimize the impact of many arthropod pests, especially blood-feeding and nuisance fly activity (Figure 2.3). Creating mounds and high places in feedlots provides dry loafing areas for livestock and effective drainage of moisture to reduce a lot of fly breeding opportunities. Routine scraping and grading over accumulated manure and wet spots in feedlots, dairies, and other confined livestock areas will also help to disrupt fly breeding activity. Keeping weeds and brush cut down will also help to allow proper air movement over confined feeding areas and prevent harborage sites where flies and mosquitoes often rest. In pastures, keeping brush and weeds cut down or burned may reduce the abundance of ticks.

Insect trapping is another form of mechanical/physical pest control. The trapping of insects can sometimes be effective in confined livestock and poultry operations to reduce fly populations. Also, trapping can be used to monitor insect populations in assessing control needs. Insect traps can employ sticky surfaces associated with light-reflective traps to mechanically catch insects or employ electrical lighting, such as "black lights," to attract insects. Electrical "fly zappers" can also be employed (Figure 2.4). These electrical devices are best used in enclosed spaces such as milk rooms and poultry processing areas and are not as effective where there are large reproducing fly populations such as at feedlots or poultry production facilities where there is a continuous accumulation of animal wastes.

Figure 2.3
Feedlot design. (Photo reproduced with permission by John B. Campbell, University of Nebraska.)

Figure 2.4
Electrical fly zapper.

Another aspect of mechanical control is that of insect exclusion. In animal product processing areas such as milk parlors and egg processing and handling, the use of screens and barriers to prevent outdoor flying insect entry can be of benefit.

BIOLOGICAL METHODS

Ectoparasites often have natural enemies that can be utilized as an integral part of a pest management program. Biological control consists of the use, exploitation, or manipulation of one life form to suppress populations of another. Pathogens, parasites, predators, and competitor species are such organisms used in biological control practices.

Numerous research studies have been made on the suitability of parasites for pest insect control. Hymenopteran parasitic wasps of fly pupae to control muscoid flies associated with confined livestock and poultry operations have been extensively studied. These parasites have proven quite successful in many livestock, poultry, and horse management systems. Several commercial biological control companies produce these parasites and offer them for sale as part of an integrated control program. By providing the proper species for release in the different geographical areas and customizing for individual animal management systems, and in proper numbers and release distribution, this approach to controlling muscoid fly populations can be very effective. Although commonly referred to as "parasites," these hymenopteran wasps actually kill their host. True parasites do not usually kill their host but obtain nourishment from their host at the host's expense. These are best referred to as parasitoids because the host is killed. Typically, a female wasp finds a fly host pupa, drills a hole into the pupa, and deposits one or more eggs. The emerging larval wasp consumes the host within the pupal case and emerges as an adult wasp. The most commonly used pupal parasitoids belong to the family Pteromalidae (Figure 2.5) and include primarily species within the genera *Muscidifurax*, *Spalangia*, and *Nasonia*.

There are several kinds of predaceous arthropods that attack pest species associated with livestock, poultry, and companion animals. Predatory beetles in the families Staphylinidae and Histeridae (Figure 2.6) usually attack the egg stage or early larval instars of muscoid filth-breeding flies. Red-legged ham beetles (*Necrobia rufipes*, family Cleridae) are also often observed as a fly predator. Earwigs in the order Dermaptera are another predatory insect that will readily feed on filth-breeding flies. They are often found in accumulating poultry waste in caged-egglayer operations. One fly species, *Hydrotaea aenescens*, commonly called the dump fly, is also considered beneficial in some situations because larvae of this species readily feed on immature stages of pest muscoid fly species. In some enclosed egglayer operations, this fly has actually displaced house flies as the primary fly species present. In this kind

Figure 2.5 (see color insert following page 132)
Pteromalid wasp. (Photo reproduced with permission by Jim Kalisch, University of Nebraska.)

Figure 2.6
Carcinops pumilio *beetle.*

of enclosed environment, they are not considered as a pest. If found breeding in outdoor areas, dump flies can be of a nuisance concern.

Certain entomopathogenic nematodes are considered beneficial. *Steinernema* spp. have shown to be promising in parasitizing fleas and ticks and some fly species. *Heterorhabditis* spp. have also shown promise for controlling house flies. Pathogenic fungi (*Beaveria* sp.) and bacteria (*Bacillus thuringiensis*) have also been extensively studied as biological control agents with some commercial products being marketed. Another aspect of biological control is habitat competition. Animal wastes often support a multitude of organisms competing for this habitat. The mass release of dung beetles (family Scarabaeidae) has proven effective in controlling manure-breeding flies, especially horn flies (*Haematobia irritans*), in warm climate regions of the United States. These beetles can destroy bovine dung pats before flies have time to complete their larval development. Most of these beetle species have been introduced from Australia and Africa. Other beetle activity in accumulated wastes in caged-egglayer operations can contribute to the drying of the habitat, making it less suitable for fly development.

Biological control is an important aspect of integrated pest management, especially with environmental concerns of chemical pesticide usage, pest resistance, and costs. Continued research on biological control of arthropods affecting animal production is of utmost importance for maintaining effective pest management programs.

CHEMICAL METHODS

Historically, many chemical agents have been used for controlling arthropods affecting livestock, poultry, and companion animals. Early compounds used were often developed by trial and error. Many of the products used were often highly toxic, containing such active ingredients as arsenic, mercury, tar, petroleum, nicotine, rotenone, and sulfur derivatives. Such chemicals have largely been replaced by pesticides that are inherently less toxic. Pesticides containing older active ingredients such as organophosphates, organochlorines, carbamates, and formanidines are still available but are being replaced by newer compounds such as pyrethroids, various kinds of insect growth regulators, and fermentation metabolites. The following are specific classes of pesticides that have products either currently being used or have been used in veterinary pest control.

Chlorinated Hydrocarbons (Organochlorines)

Chlorinated hydrocarbons have been used for controlling arthropods since the 1920s. Many of the compounds in this class are persistent in the environment and are no longer available—dichloro-diphenyl-trichloroethane (DDT) being a major example. Toxaphene and lindane, once used extensively on livestock, are also no longer available for veterinary use in the United States. Methoxychlor is one chlorinated hydrocarbon compound that still has veterinary use registration. Its mode of action is as a sodium channel modulator.

Organophosphates

Organophosphates have had wide use in veterinary pest control. These compounds are potent cholinesterase inhibitors. Some serve as quick "knockdown" insecticides such as dichlorphos, and others exhibit quick action with residual activity to continue to control target species for variable lengths of time after application. Products containing such active ingredients as malathion, chlorpyrifos, diazinon, and tetrachlorvinphos are examples of insecticides that have had extensive veterinary use. Some organophosphates have been used as systemic insecticides for controlling cattle grubs and blood-sucking arthropods. Systemics are absorbed and carried throughout the animal's system. Others have been used as feed-through treatments, passing through the digestive tract with little absorption, excreted with the feces to kill dung-breeding insects.

Carbamates

Similar in mode of action as organophosphates, anticholinesterase activity, a few carbamates have veterinary use. Carbaryl is the primary carbamate that has veterinary registrations. It is used in poultry for external parasite control and on pet animals for flea and tick control.

Formamidines

Of this class of pesticides, amitraz is the only compound used in veterinary pest control. Its mode of action is as an octopaminergic agonist. It is used as a dip or spray for the control of ticks on cattle and sheep, lice on cattle and pigs, sarcoptic mange on pigs, and demodectic mange and ticks on dogs.

Macrocyclic Lactones

Within this relative new group of pesticides appear the avermectins and milbemycins. These compounds are fermentation metabolites of the actinomycete *Streptomcyes* spp., and they act as chloride channel modulators. These chemicals have a broad spectrum of activity against arthropods and nematodes with low vertebrate toxicity. The avermectins include ivermectin, abamectin, and doramectin. The milbemycins include milbemycin, nemadectin, and moxidectin for veterinary use.

Pyrethrins

Pyrethrins are natural insecticides extracted from seed cases of the perennial plant pyrethrum (*Chrysanthemum cinergriaefolium*). Pyrethrins attack the central nervous system of insects acting as sodium channel modulators. The compound is considered a very safe insecticide for use on and around animals, is nonpersistent, and is biodegradable, rapidly breaking down on exposure to sunlight. Pyrethrins products are best used as rapid knockdown treatments with no residual activity. They are available in many kinds of formulations, including dusts, sprays, and shampoos, and often contain the synergist piperonyl butoxide. Products containing pyrethrins are used for a wide variety of pest species, including ectoparasites and flying and crawling insects.

Pyrethroids

Pyrethroids are a series of synthetic chemicals based on the chemical structure of natural pyrethrins. Although having a similar mode of action, many of the individual pyrethroid compounds have greater stability than pyrethrins with longer residual effectiveness on target pests. In many formulations, their activity is enhanced when combined with the synergist piperonyl butoxide. Pyrethroids are available as dusts, sprays, shampoos, and impregnated resin configurations, such as cattle ear tags and pet collars. Commercially available pyrethroid products contain such active ingredients as allethrin, bifenthrin, cyfluthrin, cyhalothrin, cypermethrin, deltamethrin, fenvalerate, permethrin, sumithrin, tetramethrin, and tralomethrin.

Insect Growth Regulators (IGRs)

Several products are available, known as insect growth regulators (IGRs). These materials interfere with various aspects of the growth and development of arthropods, disrupting their metamorphosis or reproduction. IGRs can be classified by their mode of action as either juvenile hormones, chitin synthesis inhibitors, or others.

Juvenile hormones are compounds produced by insects to prevent completion of metamorphosis until the insect larva is fully grown. Normal depletion of juvenile hormone triggers growth of the immature insect. Removal of juvenile hormone from young larvae induces early pupariation. This could result in prevention of adult emergence or emergence of deformed adults. Adding juvenile hormone to maturing larvae may postpone or suppress metamorphosis. Several juvenile hormone analogues have been developed. Included among them with uses on animal pests (e.g., fleas, flies) are methoprene, fenoxycarb, and pyreproxifen.

Chitin synthesis inhibitors act in preventing the proper development of the arthropod exoskeleton. Chitin is an important component in the exoskeleton structure. IGRs that prevent or inhibit chitin synthesis can be effective in killing insects at the stage when chitin synthesis is critical for growth and survival, often at egg hatch and larval molt. Such products as diflubenzuron, triflumuron, and lufenuron have registrations for fly or flea control.

Another group of IGRs includes the triazine derivatives. Cyromazine is the only commercially developed product used primarily for fly control. Cyromazine disrupts molting and pupariation not by inhibiting chitin synthesis, but somehow by acting through the hormone which controls ecdysis and the deposition of the cuticle. It is most active on muscoid flies and is used as feed additives or sprays.

Other Chemicals

There are some newer groups of chemicals showing insecticidal properties, a few of which have been commercially developed. Imidacloprid is a chlorinated analogue of nicotine and acts as a neurotoxin (acetylcholine mimic). Formulations are available for use in flea control in pets, fly control baits, and a wide variety of other pest control applications. Spinosad is another new class of insecticides with the active ingredient derived from a naturally occurring soil-dwelling bacterium *Saccharopolyspora spinosa*. The bacterium produces toxic compounds while being fermented. The mode of action of spinosad causes rapid excitation of the insect nervous system. It is used to kill a variety of pest insects but is primarily used for fly control associated with veterinary concerns. Another newer insecticide is fipronil that inhibits gamma-aminobutyric acid (GABA) receptors in invertebrates. Fipronil is used for flea and tick control in pets and has activity on several lice species.

Methods of Chemical Applications

Several application methods exist for insecticides used to control arthropods of veterinary importance. On animal, use can be by topical application (Figure 2.7), systemic application, or using feed additives or bolus treatments.

Figure 2.7
Spraying cattle with insecticide.

Premise, off-animal applications include a variety of methods from sprays to baits to dust treatments. The choice method used will vary based on the target species, host species, host management system, labor and application equipment needed, and efficiency of the treatment regime.

Topical animal application ranges from dips, wipe-ons, sprays, dusts, pour-ons, installation of insecticide-impregnated collars or ear tags, and self-application methods. Direct animal treatment using sprays or dusts requires proper application and animal-handling facilities. Most formulations used provide relatively short control effectiveness with frequent reapplication needed. With animal dips, animals are treated either in run-through dipping vats (Figure 2.8) as in the case for controlling ticks and scabies in cattle, or by hand-dipping individual animals in a dip tank (e.g., young pigs to control lice or mange). Wipe-on applications are often used for individual animal treatment such as on horses for fly control. Pour-on and spot-on treatments generally have been used for systemic treatment, but with some of the pyrethroids and newer compounds like spinosad, these application methods work topically.

Extensive use had been made of resin insecticide–impregnated pet collars and cattle ear tags. Insecticide pet collars are used for flea and tick control, and cattle insecticide ear tags (Figure 2.9) are used primarily for pasture fly control (e.g., face flies, horn flies). These devices are designed for longer-term control, often with one application during the pest season. These devices work by the insecticide continually being released from the device over time and the insecticide being spread over the skin and hair coat of the host animal.

The use of self-application devices allows for self-treatment by animals. In range cattle, insecticide dust bags (Figure 2.10) or backrubbers (Figure 2.11) can be employed to control face flies and horn flies during warmer months and cattle lice during colder months. For maximum effectiveness, however, they need to be placed in forced-use situations to assure animal usage, such as in gaps between pastures and entrances to water sources and mineral supplements. Also, these devices should be configured so animals get adequate coverage of the insecticide to control target pests (e.g., flies on the face and back). Dust bags and backrubbers need frequent maintenance to assure adequate supplies of insecticide, proper insecticide release, and any needed repairs or adjustment. Also, these devices should be protected from direct exposure to rainfall by using either waterproof materials or some type of roof protection.

Systemic insecticides work by being absorbed through the host skin and systemically moving through the animal's system. Originally used for controlling cattle grub infestations in cattle, they have use on a variety of internal arthropod parasites and blood-feeding arthropods of veterinary importance. Formulations used as systemic treatments range from injectables to oral and topical application methods. Systemic treatments are of particular value for pest arthropods that spend all or a considerable portion of their lifecycle on the host or are dependent on the host for frequent blood-feeding. Some systemic compounds will kill arthropods and other internal parasites and

Figure 2.8
Cattle dip vat.

Figure 2.9
Installing cattle insecticide-impregnated ear tags.

Figure 2.10
Cattle insecticide dust bag.

Figure 2.11
Cattle insecticide backrubber.

are commonly referred to as parasiticides. Pour-on and spot-on treatments (Figure 2.12) are common methods of delivery of many systemic compounds. Some compounds such as the avermectins are applied by injection under the skin or into the tissue of the host. These methods are convenient and have in many cases replaced dips and sprays for those purposes. With production livestock though, adequate animal handling facilities are needed for proper treatment application.

Oral applications by use of feed-additives (Figure 2.13) and bolus treatment (Figure 2.14) can either be for systemic treatment or nonsystemic use. Nonsystemic products are used primarily to pass through the digestive tract of the host unchanged and are then available in the feces to control developing fly larvae (e.g., house flies, face flies, and horn flies). Bolus treatments are designed to allow a sustained release of the insecticide over several weeks, as the bolus stays in the stomach of the host. Boluses are used primarily in ruminant animals.

Various methods can be employed for chemical control of arthropod pests off host animals, including sprays, baits, dusts, and larvicides. Insecticide spray applications can be made either as quick knockdown space sprays (e.g., for flies, mosquitoes, etc.) or as residual spray treatments. Space sprays consisting of small droplets, mists, or fogs are often an effective means of quick knockdown of flying insects such as flies and mosquitoes. They are best used in confinement livestock and poultry production facilities where large numbers of adults flies sometimes are present. Equipment available can either be

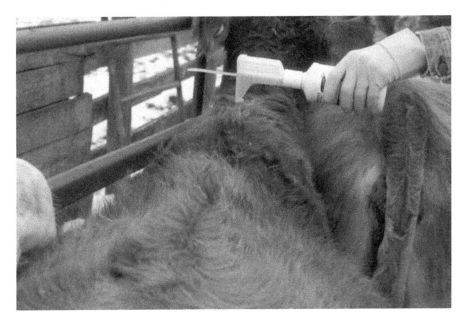

Figure 2.12
Systemic insecticide spot-on treatment.

Figure 2.13
Cattle feed-additive treatment. (Photo reproduced with permission by John B. Campbell, University of Nebraska.)

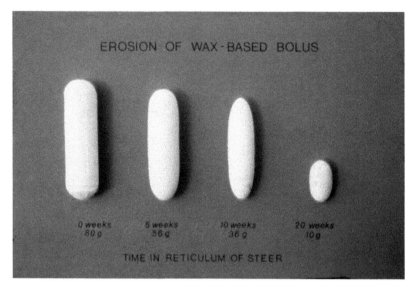

Figure 2.14
Cattle insecticide bolus.

portable (e.g., backpack sprayers, truck-mounted sprayers) (Figure 2.15 and Figure 2.16) or built into a structure (e.g., pipelines equipped with numerous spray nozzles). Portable equipment is most often preferred for ease in making application throughout different facilities and in readily being able to change insecticide formulations as needed. The preferred equipment to use should produce small enough droplets to move through the space treated and make contact with the target insects. Ultra-low-volume cold aerosol application equipment as used in most mosquito control operations is preferred. Droplet size is ideal, and this method typically uses insecticides that are ready to use and do not have to be diluted. Synergized pyrethrins and pryethroids are most commonly used. Insecticide mist application usually has larger droplet sizes and is sometimes used to spray over animals or used with wall- or ceiling-mounted sprayers. Thermal fogging is accomplished with an oil and insecticide mixture applied in a thermal fogging machine. This produces a very small droplet size that appears as a "smoke." Most such sprayers are portable handheld or truck-mounted sprayers.

Residual sprays are applied to surfaces (Figure 2.17) where pest arthropods frequent (e.g., fly resting places, crack and crevice arthropod harborage areas, etc.). Residual sprays are most frequently used in confinement livestock and poultry operations, horse facilities, pet kennels, and so forth, where there are defined surfaces where pest species are found resting. Flies tend to rest on vertical surfaces and higher up in buildings, especially at night. Residual sprays applied to these areas are normally effective for a week or longer depending on moisture, temperature, sunlight exposure, and dust buildup. The major concern with using residual insecticide treatments is pest resistance

Figure 2.15
Insecticide backpack sprayer.

Figure 2.16
Truck-mounted ULV (ultra-low-volume) insecticide sprayer.

Figure 2.17
Residual insecticide spray treatment. (Photo reproduced with permission by John B. Campbell, University of Nebraska.)

to the compounds used. Typically, most residual sprays consist of pyrethroids and a few organophosphate insecticides. Insects, especially house flies, have shown the ability to develop a genetic resistance to the repeated use of these residual insecticides.

Some insecticide sprays and dust products can be used as larvicides and are applied directly on fly breeding areas. They are most frequently used in confined livestock and poultry operations where animal wastes accumulate and cannot be removed at frequent intervals. Cyromazine has registrations for direct spray application and as a feed additive in poultry feed to control flies breeding in poultry wastes. Larvicide application should be limited, because as with most insecticides being used, fly resistance has been shown to occur with extensive and repeated use. Also, many larvicides used are detrimental to nontarget organisms, including those that aid in the biological control of flies.

The use of insecticide baits is another means of controlling insect pests, primarily house flies. The most commonly used baits are granulated or are in extruded pellet form (Figure 2.18). They contain both toxicant and attractant to lure flies. They can be used as scatter baits, broadcast on walkways or other dry areas where flies can be attracted to and other animals cannot get to. They can also be placed in protective bait stations out of animal reach. Some bait products can also be mixed with water to form a slurry that can be brushed onto a fly resting surface. Label directions on any given product will provide how it can be used.

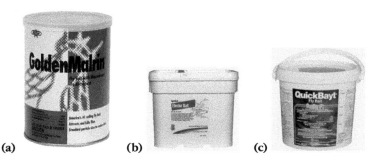

Figure 2.18
Insecticide fly bait. (a) Golden Malrin (methomyl); (b) Elector Bait (spinosod); (c) QuickBayt (imidacloprid).

PESTICIDE SAFETY

When using any pesticide product, proper precautions must be taken. Although most pesticides for veterinary use are considered relatively safe, if misused they can lead to human or animal poisoning. It is imperative to read and follow all instructions of the product label before using any pesticide. The label not only provides mixing instructions and directions for use, but will spell out any use restrictions, animal slaughter time intervals after treatment, animal species and management systems that can treated, what protective gear and clothing should be used, and any other pertinent information about the use of the product. Manufacturer contact information is also available on the label in case specific questions about the product need to be answered. It is also imperative to use only products for veterinary use when controlling animal pests. Some insecticides have both agronomic and veterinary formulations. In many cases, the inert "carrier" ingredients used for crop application may be different than those for veterinary use. Use the product chosen only for the labeled use intended. It should also be noted that the pesticide label is a legal document, and use inconsistent with label directions is illegal.

CHAPTER 3

FLIES (DIPTERA)

TABLE OF CONTENTS

One of the largest and most diverse orders of insects is the true flies, Diptera. There are many species of Diptera that are of significant veterinary importance (Table 3.1). Diptera affecting domestic animals range from small biting flies to larger species, wingless ectoparasites, larvae that invade animal tissue, to nonbiting flies that can be of significant nuisance concern. Many fly-borne diseases affect domestic animals. The extent of flies associated with both human and animal diseases can be reviewed in the two-volume treatise *Flies and Disease* by Greenberg (1971, 1973) and in Mullen and Durden's *Medical and Veterinary Entomology* (2002).

Historically, the order Diptera has been divided into three distinct suborders: Nematocera, Brachycera, and Cyclorrhapha. More recently, classification has been adopted to include two suborders: Nematocera and Brachycera.

The Nematocera are considered to be the more primitive Diptera and include smaller flies with long/narrow wings and long antennae composed of several segments. The larvae of Nematocera have well-developed head capsules. Among the more notable families in this suborder of veterinary importance include the Ceratopogonidae (biting midges), Culicidae (mosquitoes), Simuliidae (black flies), and Psychodidae (sand flies, moth flies).

Brachycera of veterinary importance include two infra-orders: Tabanomorpha and Muscomorpha. The Tabanomorpha include Tabanidae (horse flies, deer flies) and Stratiomyidae (soldier flies) which are of veterinary importance. Flies in this group are usually larger, more robust flies. Larvae have a head capsule that is relatively incomplete and often retracted into a fleshy body.

The Muscomorpha is a huge infraorder and include a diverse group of fly families of veterinary importance. Of most concern are Muscidae (house fly, stable fly, horn fly, others), Glossinidae (tsetse flies), Hypoboscidae (louse flies), and the superfamily Oestroidea that includes Calliphoridae (blow flies), Sarcophagidae (flesh flies), Gasterophilidae (horse bots), Hypodermatidae (cattle grubs), Cuterebridae (rodent bots), and Oestridae (sheep nose bot, others). Adults in this group have three-segmented aristate antennae. The wing veination is somewhat reduced as compared to the Nematocera and Tabanomorpha. Larvae have no definite head capsule but only a cephalopharyngeal skeleton (mouth hooks) usually seen through their translucent cuticle. The larvae are generally maggot-like or grub-like in appearance.

Covered in this chapter of this diverse group of insects will be the biting and nonbiting flies that are of importance as ectoparasites and/or nuisance flies. Chapter 4 will discuss myiasis-producing flies that invade animal tissue.

Table 3.1
Flies of Veterinary Importance (Order Diptera)

Species	Common Name
Suborder Nematocera	
Family Ceratopogonidae	Biting midges
Culicoides sp.	
Leptoconops sp.	
Family Culicidae	Mosquitoes
Subfamily Anophelinae	
Anopheles sp.	
Subfamily Culicinae	
Aedes sp.	
Ochlerotatus sp.	
Psorophora sp.	
Culex sp.	
Culiseta sp.	
Coquillettidia sp.	
Others	
Family Psychodidae	Moth flies
Lutzomia sp.	
Phlebotamus sp.	
Psychoda sp.	
Telmatoscopus sp.	
Family Simuliidae	Black flies
Simulium sp.	
Suborder Brachycera	
Infraorder Tabanomorpha	
Family Tabanidae	Horse and deer flies
Family Stratiomyidae	
Hermetia illucens (Linnaeus)	Black soldier fly
Infraorder Muscomorpha	
Family Syrphidae	
Eristalis tenax (Linnaeus)	Rat-tailed maggot
Family Piophilidae	Cheese skipper flies
Family Chloropidae	
Hippelates sp.	Eye gnats
Family Muscidae	
Stomoxys calcitrans (Linnaeus)	Stable fly
Haematobia irritans (Linnaeus)	Horn fly
Musca autumnalis (De Geer)	Face fly

Continued

Table 3.1 (Continued)
Flies of Veterinary Importance (Order Diptera)

Species	Common Name
M. domestica (Linnaeus)	House fly
Fannia canicularis (Linnaeus)	Little house fly
F. scalaris (Fabricus)	Latrine fly
Muscina stabulans (Fallen)	False stable fly
Hydrotaea leucostoma (Wiedamann)	Black garbage fly
H. aenescens (Wiedamann)	Black dump fly
Family Glossinidae	
Glossina sp.	Tsetse flies
Family Calliphoridae	
Cochliomyia hominivorax (Coquerel)	Screwworm fly
C. macillaria (Fabricus)	Secondary screwworm
Lucilia sp.—Green bottle flies	Green bottle flies
Calliphora sp.—Blue bottle flies	Blue bottle flies
Phormia regina (Meigen)	Black blow fly
Chrysomyia rufifacies (Macquart)	Hairy maggot blow fly
Others	
Family Sarcophagidae	Flesh flies
Wohlfahrtia vigil (Walker)	
W. magnifica (Schiner)	
Sarcophaga sp.	
Family Oestridae	
Subfamily Hypodermatinae	
Hypoderma lineatum (de Villers)	Common cattle grub
H. bovis (Linnaeus)	Northern cattle grub
Oedemagena tarandi (Linnaeus)	Caribou warble fly
Subfamily Gasterophilinae	
Gasterophilus intestinalis (De Geer)	Horse bot fly
G. nasalis (Linnaeus)	Throat bot fly
G. haemorrhoidalis (Linnaeus)	Nose bot fly
Subfamily Oestrinae	
Oestrus ovis (Linnaeus)	Sheep bot fly
Subfamily Cuterebrinae	
Cuterebra sp.	Rodent bot flies
Dermatobia hominus (Linnaeus)	Human bot fly
Subfamily Cephenemyiinae	
Cephenemyia sp.	Deer bot flies
Family Hippoboscidae	
Melophagus ovinus (Linnaeus)	Sheep ked

BITING FLIES

Grouped here are those flies that don't literally "bite" but actually feed on blood by piercing the skin of a host with piercing–sucking mouthparts, or other adaptations for breaking through the skin. These flies are of the most economically important groups of arthropods associated with animal production and health. Biting flies contribute to more than 50% of the estimated annual losses to livestock production by arthropods from direct loss attributable to exsanguination and animal annoyance to transmission of pathogenic organisms and secondary infection.

Suborder Nematocera

Mosquitoes (Culicidae) (Figure 3.1)

Geographic Distribution: Worldwide

Veterinary Importance: Over 3000 species of mosquitoes occur worldwide, several of which occur in North America. Two subfamilies are of primary medical and veterinary importance: Anophelinae (e.g., *Anopheles*) and Culicinae (e.g., *Aedes, Ochlerotatus, Psorphora, Culex, Culiseta, Coquillettidia*, and others). Mosquitoes are a direct cause of annoyance and blood loss to livestock and companion animals and serve as vectors of a multitude of disease agents. They can disrupt normal animal behavior and grazing. Increased scratching from mosquito bites can result in skin abrasions, hair loss, and secondary

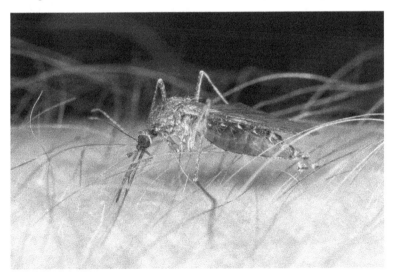

Figure 3.1
Adult Culex *sp. mosquito. (Photo reproduced with permission by Jim Kalisch, University of Nebraska, Lincoln.)*

infection. Mosquito-borne diseases affecting domestic animals range from viruses (e.g., encephalitis, West Nile virus, Rift Valley fever) to nematodes (e.g., dog heartworm).

Hosts: Most domestic animals

Description: Adult mosquitoes can be distinguished from other Diptera by several characteristics (Figure 3.2). They have long, many segmented antennae,

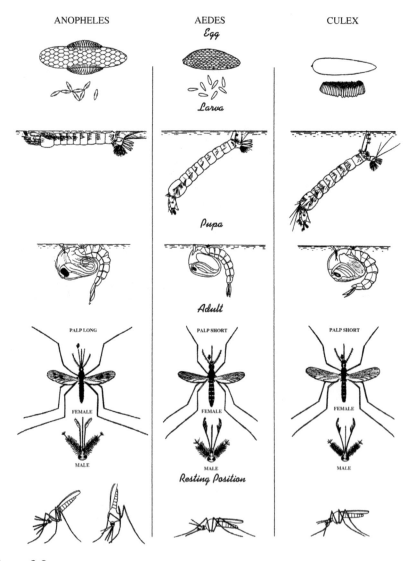

Figure 3.2
Characteristics of anopheline and culicine mosquitoes. (Illustration from the Communicable Disease Center, U.S. Department of Health, Education, and Welfare.)

a long proboscis for blood sucking (females), and scales on their wing fringes and veins. Adult males can be distinguished from females by their "bushy" antennae. Also, adult males do not feed on blood.

Differences occur between the appearance of anopheline (e.g., *Anopheles* sp.) and culicine (e.g., *Aedes, Culex*, other sp.) mosquitoes. The palps of female anopheline mosquitoes are as long as the proboscis, whereas the palps of female culicine mosquitoes are short, and only about one quarter the length of the proboscis. When resting on a flat surface, anopheline mosquitoes rest with the proboscis, head, thorax, and abdomen in a straight line at an angle to the surface. Culicine adults rest flat to the surface. Also, anopheline mosquitoes frequently have distinct scale patterns on their wings giving them a spotted or mottled appearance.

All mosquito larvae are aquatic and commonly called "wigglers." They have four growth stages, or instars. They have a distinct head capsule and have simple or branched tufts of hair along the body. In culicines there is a prominent siphon tube for breathing at the water surface. This is lacking in anopheline larvae. Pupal mosquitoes, commonly called "tumblers," are also found in water. They are comma shaped, with the head and thorax fused forming a cephalothorax with the abdomen curled underneath. There are two prominent respiratory trumpets on the thorax by which the pupa breathes at the water surface. On the last abdominal segment, there are two broad paddles to propel the pupa through the water when disturbed.

Biology/Behavior: The lifecycles and behavior of different mosquito species vary considerably (Figure 3.3). Anopheline mosquitoes lay eggs singly directly on the water surface. These eggs are equipped with air-filled compartments on each side that serve as floats. Depending on water temperature, these eggs usually hatch 2 to 3 days after being deposited. Some culicine mosquitoes (e.g., *Aedes, Ochlerotatus, Psorophora*) attach their eggs individually on a substrate that later will be covered with water to initiate egg hatch. Such eggs can often remain viable for up to 3 years if conditions are not suitable for egg hatching. Other culicine mosquitoes (e.g., *Culex, Culiseta, Coquillettidia*) lay their eggs in clumps of several eggs (100 to 300) in boat-like rafts on the water surface in which larvae will emerge in 2 to 3 days or longer, depending on water temperature.

All mosquito larvae are aquatic and develop in a wide variety of habitats. Habitats range from permanent ponds, to natural wetlands, to marshes, to areas prone to flooding, to temporary pools of water left by rainfall, to tree holes and other natural water-holding cavities, to a variety of artificial containers (e.g., discarded tires, clogged gutters, open trash cans, water troughs, etc.) (Figure 3.4). Depending on species, temperature, and habitat, larvae will take from 4 to 5 days to 2 to 3 weeks or longer to develop. Once the larvae pupate, the pupal stage will last another 2 to 7 days.

Adult mosquitoes emerge from pupal cases in the water and will take flight a few minutes after emergence when their cuticles sclerotize. Mating may

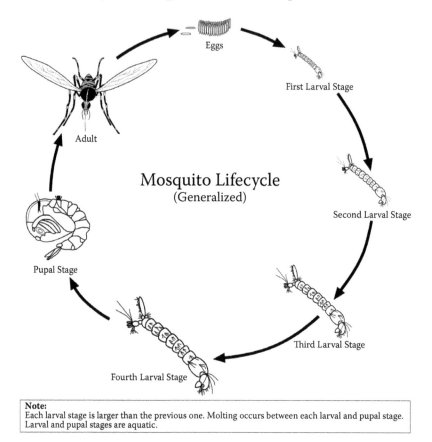

Note:
Each larval stage is larger than the previous one. Molting occurs between each larval and pupal stage. Larval and pupal stages are aquatic.

Figure 3.3
Mosquito lifecycle. (Illustration from Scott Charlesworth, Purdue University.)

occur within 1 day to several days after emergence. Only females feed on blood (for egg development); males get their nourishment from nectar and other sources.

Mosquitoes vary considerably in their host preferences. Some species prefer mammalian blood, others prefer avian hosts, and others will feed almost exclusively on amphibian or reptilian blood. Flight also varies considerably, with some species staying within 1.6 km of emergence and other species seeking hosts as far as 16 to 32 km away. Females of many species take multiple blood meals and produce multiple batches of eggs. Some species, such as *Aedes*, *Ochlerotatus*, and *Psorophora*, overwinter in the egg stage on dry substrates. Others, such as *Anopheles* and *Culex*, overwinter as fertilized females in temperate regions.

Black Flies (Simuliidae) (Figure 3.5)

Geographic Distribution: Worldwide

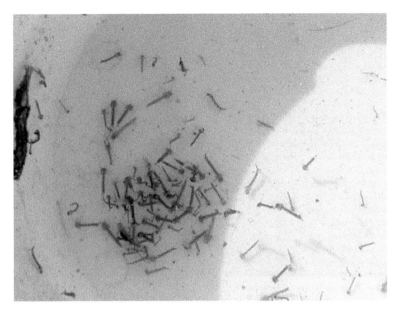

Figure 3.4
Mosquito larvae in mosquito dipper taken from a water trough.

Figure 3.5
Black fly. (Photo courtesy of Oklahoma State University.)

Figure 3.6
Black flies feeding in horse ear. (Photo courtesy of University of California.)

Veterinary Importance: Black flies, also known as buffalo gnats and turkey gnats, are of veterinary importance as blood feeders, feeding on poultry, cattle, swine, sheep, goats, horses, and other animals. They are usually found feeding on the legs, abdomen, head, and in ears. One species, *Simulium vittatum*, frequently is found feeding in the ears of horses, cattle, swine, sheep, and other animals causing pain, scabbing, and irritation (Figure 3.6). Another species, *S. meridionale*, known as the turkey gnat, frequently feeds on the comb and wattles of poultry. This species is found in the Mississippi Valley.

When large numbers of black flies are present, attack by black flies has sometimes led to animal death. Many deaths can be attributed to acute toxemia of black fly bites. Even at low populations, black fly bites can be painful and cause considerable animal disturbance and reduced productivity. Livestock under constant attack often will stampede, crushing young animals and getting hurt by running into fences and structures. Large numbers of black flies feeding can also result in anemia and exsanguination. Black flies have also been associated with animal disease transmission, being vectors of the filarial nematodes (*Onchocerca*) causing bovine onchoceriasis and protozoans (*Leucocytozoon*) causing leucocytozoonosis in poultry.

Hosts: A wide range of domestic and wild mammals and birds

Description: Adult black flies are small and robust (~3 to 6 mm long) with a characteristic humped thorax and relatively short wings. They have bead-like antennae. The palps extend beyond the proboscis. Coloration varies from black, gray, to a dark yellow or orange.

Figure 3.7
Black fly larvae, Simulium vittatum. *(Photo reproduced with permission by Jim Kalisch, University of Nebraska, Lincoln.)*

Immature black flies develop in clean, swift-flowing streams. Most black fly larvae possess a pair of mouth brushes on a well-sclerotized head capsule used for filtering food particles in their aquatic habitat. They have an elongate body with a thoracic proleg and terminal proleg used to adhere to underwater substrates (Figure 3.7). Black fly pupae are housed in silk cocoons that attach to underwater substrates.

Biology/Behavior: Immature black flies develop primarily in shallow, clean, swift-flowing streams (Figure 3.8). Larvae attach to submergent substrates such as stones and vegetation. They are filter feeders, feeding on suspended organic matter from water current passing by. There are usually 6 to 11 larval instars. The duration of the larval stage varies from about 1 week up to a year, depending on species, stream temperature and condition, and food availability. Final-stage larvae pupate in silk cocoons that are spun on underwater substrates. Newly emerged adults rise to the water surface. Adults generally live less than a month. Adults of most species are prevalent during late spring and early summer and often are present in large swarms where mating occurs. Female black flies normally bite in the daytime, but more frequently in early morning and at dusk. Females require blood for egg development. Eggs are laid in groups of 150 to 600 in or near a water source. Male black flies do not bite but obtain nourishment from sugar sources.

Biting Midges (Ceratopogonidae) (Figure 3.9)

Common Names: Punkies, no-see-ums, biting gnats

43

Figure 3.8
Swift flowing stream black fly habitat. (Photo courtesy of Jesse Hobbs, Centers for Disease Control and Prevention.)

Figure 3.9
Culicoides *sp. biting midge. (Photo courtesy of University of California.)*

Geographic Distribution: Worldwide in temperate and tropical regions

Veterinary Importance: Adult biting midges are vicious and persistent biters. Some 600 species of ceratopogonids occur in North America. Those in the genera *Culicoides* and *Leptoconops* are of most importance to domestic animals. Biting midges are known to vector several disease agents. Most prevalent are viruses causing bluetongue in sheep and cattle, epizootic hemorrhagic disease in cattle and wild ruminants (e.g., deer), bovine ephemeral fever, and African horse sickness in equines (e.g., horses, donkeys, mules). They also are known to vector nematodes responsible for equine onchoceriasis and blood protozoans in poultry. In certain horse breeds, biting midges can also cause a skin reaction known as equine allergic dermatitis.

Hosts: Wide range of domestic and wild animals

Description: Adult biting midges are very small flies ranging from 1 to 4 mm in length. Their wings are distinctly mottled, and at rest the wings are folded over each other and held flat over the abdomen. Ceratopogonid larvae are long and slender (eel-like) ranging from 2 to 5 mm in length at maturity. They have a distinct brownish head capsule and creamy-white body. Pupae are brownish in color, with a pair of relatively short prothoracic respiratory horns at the anterior end. They use these to hang at the water surface to breath air.

Biology/Behavior: Adult female ceratopogonids feed on blood for egg development. However, some species are autogenous during their first gonotrophic cycle without feeding on blood. Subsequent egg development usually requires a blood meal. Adult males feed primarily on nectar. Many species feed mostly on mammals, and others prefer to feed on birds, reptiles, or amphibians. Preferred feeding activity for most species is in early morning and at dusk, but some will feed during the day. Biting midges usually stay in close proximity to their breeding sites. Eggs are deposited in batches of 25 to more than 400 eggs on moist surfaces at larval breeding habitats. The larvae are aquatic and occur in a wide variety of aquatic and semiaquatic habitats. Habitats range from edges of lakes and streams, marshes, swamps, wetlands, shallow margins of ponds, water-filled tree holes and cavities, axils of plants, and other aquatic habitats. Some species will even be found breeding in moist soil, ground litter, and animal wastes. Larvae are frequently encountered around muddy edges of their aquatic habitat where they feed on a variety of microorganisms and organic matter. Ceratopogonid larvae have a distinct serpentine movement as they move through water. Larvae have four instars with developmental time ranging from 2 weeks to a year or longer depending on species and geographic location. Most species overwinter as larvae. Pupation generally occurs near the water surface where the pupae can penetrate the surface with their prothoracic respirator horns. Most species of ceratopogonids have one to two generations per year.

Suborder Brachycera (Infraorder Tabanomorpha)

Horse Flies/Deer Flies (Tabanidae) (Figure 3.10 and Figure 3.11)

Geographic Distribution: Worldwide

Veterinary Importance: Horse flies and deer flies are very common pests of livestock and companion animals, especially cattle and horses kept outdoors. Over 300 species are known to occur in North America. Most are larger, robust

Figure 3.10 (see color insert following page 132)
Horse fly, Tabanus atratus. *(Photo reproduced with permission by Jim Kalisch, University of Nebraska, Lincoln.)*

Figure 3.11
Deer fly, Chrysops vittatus. *(Photo courtesy of Patrick Jones, Purdue University.)*

Figure 3.12
Horse fly larva. (Photo courtesy of Patrick Jones, Purdue University.)

flies as compared to other fly groups. They are known for their painful bites and persistent nuisance to animals. Their mouthparts are stout and bladelike and are capable of inflicting deep, painful, bleeding wounds in which they lap up blood that pools at the surface. Heavy attack by tabanids can result in reduced weight gains in cattle, reduced milk production, and direct skin irritation. Tabanids are capable of vectoring several disease agents of veterinary importance including viruses (equine infectious anemia, bovine leukemia, hog cholera), bacteria/rickettsia (*Anaplasma marginale, Francisella tularenisis, Bacillus anthracis*), protozoa (*Trypanosoma* sp.), and nematodes (*Elaeophora* sp.).

Hosts: Primarily cattle and horses; less common on other mammals

Description: Tabanid adults are stout, large flies ranging in size from 6 to 30 mm. Deer flies are generally smaller than horse flies. They have large prominent eyes. The mouthparts of adult females are stout and bladelike to cut the skin of their host and with sponging labella to lap up pooled blood. Males, with less-developed mouthparts, do not take blood meals.

Tabanid larvae are cylindrical and range from 15 to 30 mm long when mature (Figure 3.12). The head capsule is incomplete and partially sclerotized. On the abdominal segments, there are three to four pairs of fleshy prolegs. Larvae breathe by way of a posterior siphon that varies in length between species. Tabanid pupae are tan to brown in color with eyes, legs, and wing pads visible. They have a fringe of spines on several of the abdominal segments and three to four pairs of caudal projections (pupal asters).

Biology/Behavior: Tabanid eggs are deposited in masses from 100 to 1000 in a single mass on vegetation overhanging aquatic/semiaquatic habitats (Figure 3.13). Eggs normally hatch in 5 to 12 days. Larvae drop to the water or mud where they feed on organic debris or prey on other aquatic life. They will prey on a variety of invertebrates from aquatic insects to annelids. Tabanid

Figure 3.13
Tabanid eggs on vegetation.

larvae are found in a wide variety of aquatic/semiaquatic habitats. They can be found along margins of ponds, creeks, streams, marshes, and wetlands. Some species may be found on the bottom of ponds and lakes, and under rocks and debris in streams. Others may be found in more terrestrial habitats such as under moist forest litter.

Tabanids will undergo 6 to 13 larval molts with varied developmental time depending on species and environmental condition. They overwinter as larvae. Mature larvae usually pupate in the spring. Pupation normally occurs above the waterline in drier spots and lasts from 1 to 3 weeks. Most temperate species have one generation per year with some southern species completing two or more generations per year. Other species, especially larger species, such as *Tabanus atratus*, may take 2 to 3 years to complete their lifecycle.

Only adult female tabanids feed on blood. Males feed on nectar, honeydew, and other plant sugars. Tabanids mate soon after emergence while in flight. Females normally are diurnal feeders, being most active on warm, sunny, calm days. Their feeding activity is affected by changes in environmental condition, especially temperature and barometric pressure changes. Tabanids are strong fliers and can travel several kilometers in search of a blood meal. A given species may be abundant for 4 to 6 weeks, but because of a multitude of species found in any given locale, livestock and equine are usually bothered throughout the warmer months.

Suborder Brachycera (Infraorder Muscomorpha)

Stable Fly (Muscidae) (Figure 3.14)

Species Name: *Stomoxys calcitrans* (Linnaeus)

Figure 3.14
Stable fly adult, Stomoxys calcitrans. *(Photo reproduced with permission by Jim Kalisch, University of Nebraska, Lincoln.)*

Geographic Distribution: Worldwide

Veterinary Importance: Stable flies are very annoying pests of livestock and domestic animals, inflicting painful bites. They favor feeding on lower portions of an animal, on the legs and undersides. On small hosts (e.g., small ruminants, dogs), they will also be seen on the head and ears, in which visible lesions can develop from stable fly bites. Of primary concern are blood loss and animal disturbance. Stable fly feeding activity can result in reduced weight gains and reduced milk production. Animals being attacked will stomp and kick and often bunch together to avoid being bitten. They are persistent and strong fliers, following host animals for several kilometers. Stable fly migration studies have indicated they can travel over several states.

Although stable flies are not considered an important vector of animal pathogens, they have been implicated as vectors of retroviruses (equine infectious anemia in horses, bovine leucosis in cattle), nematodes (*Habaronema microstoma* that causes habronemiasis in horses, and a blood parasite *Eperythrozoon suis* in swine).

Hosts: Wide range of domestic and wild animals

Description: Adult stable flies are 5 to 8 mm long and gray in color, with four longitudinal dark stripes on the thorax and "checkerboard" ventral abdominal markings. Their blood-feeding proboscis is readily visible as it projects out from the head. Mature larvae are around 10 mm long and creamy white in color. The larvae can be distinguished from house fly maggots by the size and shape of the posterior spiracles. Pupae are 6 to 7 mm long and dark brown in color when hardened.

Biology/Behavior: Both adult male and female stable flies are blood feeders. Individual stable flies typically feed once per day, feeding for up to 2 to 5 min. Engorged flies are sluggish and can be found on nearby walls, fences, and vegetation digesting their blood meal. After mating, females require additional blood meals prior to oviposition.

Eggs are deposited in wet, organic materials such as straw, litter, and manure mixed with straw or other bedding. They can be found breeding at the base of silage piles, in hay bales (especially in and under round bales), in grass clippings, in dead vegetation around flooded areas and beaches, and in poorly managed compost piles. They are seldom found in pure manure sources. Individual females can lay from 500 to 1000 eggs in a lifetime in batches of 25 to 50 eggs at a time over a 4- to 6-week period. Eggs hatch in 1 to 3 days. Larval development through three instar growth stages generally takes from 2 to 3 weeks, depending on moisture, temperature, and food availability. The pupal period ranges from 1 to 4 weeks varying with temperature. A complete lifecycle in warmer months will occur in 3 to 4 weeks. Stable flies usually overwinter as larvae or pupae protected under their habitat cover.

Horn Fly (Muscidae) (Figure 3.15)

Species Name: *Haematobia irritans* (Linnaeus)

Figure 3.15
Horn fly adult, Haematobia irritans. *(Photo reproduced with permission by Jim Kalisch, University of Nebraska, Lincoln.)*

Geographic Distribution: In essentially all cattle-producing countries world-wide, including Northern Africa, Europe, central Asia, North America, Central America, and South America. A subspecies, the buffalo fly (*H. irritans exigua*), is found in Australia and the Pacific region.

Veterinary Importance: The horn fly is a severe pest of cattle and is considered the major pastured fly pest of cattle in North America. They were introduced into North America from Europe in the mid-1880s, being first found in New Jersey in 1887. They rapidly spread across the country and by 1898 were found to have reached the Rocky Mountains.

Both sexes of horn flies feed on cattle several times a day (from 20 to 30 times). Numbers on cattle ranging from 1000 to 4000 or more are common during mid to late summer (up to 10,000 horn flies have been observed on bulls) (Figure 3.16). These infestation loads account for significant blood loss to the host animal. Pain and annoyance of horn fly infestation also affects animal behavior and grazing time. Reduced weight gain and reduced feed efficiency can result from heavy infestation. As few as 200 horn flies per animal have been observed as an economic threshold level.

Horn flies have been shown to be involved in the transmission of the nematode *Stephanofilaria stilesi* that causes a granular dermatitis on the skin of cattle. This nematode occurs in North America, primarily in the western United States and Canada.

Hosts: Almost exclusively on cattle

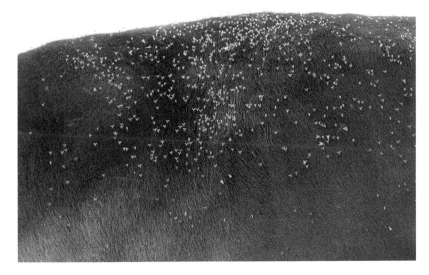

Figure 3.16
Horn flies on cow.

Description: Adult horn flies are relatively small gray-black flies, 3 to 5 mm long, with a piercing–sucking proboscis. Horn fly eggs are reddish-brown in color. The larvae (maggots) are smaller than other muscoid flies, as are the brown pupae.

Biology/Behavior: Adult horn flies are considered as continuous obligatory parasites on pastured cattle, with female flies only leaving the host animal briefly to lay eggs. Mating occurs on the host. Adults usually congregate on the back, withers, and around the head. On hot, sunny days, they often move to the underside of their host. Horn flies derive their name from observations seeing them clustering around the base of horns on cattle. While feeding, horn flies on cattle usually face in a head-downward position.

Female horn flies oviposit their eggs in fresh bovine manure in pastures, shortly after it is dropped by the animal. In her lifetime, a single horn fly female will produce 300 to 400 eggs in batches of 20 to 30 eggs. Eggs will hatch in 1 to 2 days, depending on temperature. The larvae develop within the dung pat, taking 3 to 10 days to develop through three instar growth stages. Pupation occurs within or just below the dung pat and usually takes 6 to 8 days to develop before emerging as adults. The entire lifecycle can be completed within 10 to 14 days, depending on climatic conditions and nutrient quality of the manure. In northern areas horn flies overwinter in the pupal stage, but they will reproduce the entire year in warmer climates.

Horn fly abundance will vary by region and time of year. In temperate climates, there is usually a single population peak in midsummer. In warmer climates, there is usually an early and late peak, with lower populations occurring in hotter parts of the summer.

Sheep Ked (Hippoboscidae) (Figure 3.17)

Species Name: *Melophagus ovinus* (Linnaeus)

Geographic Distribution: Worldwide, except in tropical areas where it is confined to upper highland regions.

Veterinary Importance: The sheep ked is a unique wingless fly adapted as a continuous ectoparasite spending its entire life on sheep. It is considered the most economically important ectoparasite of sheep in North America. Both females and males feed on blood and can cause considerable irritation to sheep. Animals may scratch and rub and often damage their wool coat. Heavier infestations lead to wool loss and a condition of the skin called "cockle," producing a pitting, ridging appearance as a result of an allergic response to ked bites. This can cause significant damage to the hide and downgrades the leather value. Fecal staining of the wool may also occur which reduces its value.

Hosts: Sheep and occasionally goats

Description: Sheep keds, both females and males, are wingless flies, even lacking halteres. Adults are 5 to 8 mm long. They are leathery in appearance,

Figure 3.17
Sheep ked adult, Melophagus ovinus. *(Photo reproduced with permission by Jim Kalisch, University of Nebraska, Lincoln.)*

brownish-red in color, dorsoventrally flattened, and look somewhat tick-like. They are often called sheep ticks.

Biology/Behavior: Sheep keds have a most unusual lifecycle, well adapted for being a continuous ectoparasite on sheep. Both females and males feed on blood, taking a blood meal every 24 to 36 hours, normally feeding for 5 to 10 minutes at a time. The female ked produces only one offspring at a time. The emerged larva develops within a uterine pouch inside the female. In this pouch, the larva is fed by a pair of accessory glands (milk glands) and undergoes two molts. When it molts into the third instar, it is extruded by the female and is ready to pupate. This kind of reproduction is known as larviposition in which larvae, not eggs, are deposited. Upon emerging, the third instar larva's integument hardens very quickly, forming a reddish-brown barrel-shaped puparium that is attached to the wool of the host. The duration of the pupal stage lasts from 18 to 40 days, depending on temperature, time of the year, and location on the host. Eclosed female sheep keds mate within a day of eclosion and reach sexual maturity in 14 to 30 days. A female sheep ked normally lives for about 4 months and produces 10 to 15 offspring. Males live approximately 2 to 3 months.

Fluctuation in sheep ked numbers on a host varies by time of year. Populations tend to be higher from winter to early spring and lowest during the summer months. They are often difficult to find in the wool at low population levels. Keds tend to be found in higher numbers in younger animals. Newborn lambs become infested with keds directly from their mothers soon after birth. Within a flock, infestations between animals occur when sheep keds move to the outer edges of fleece in response to increasing air

Figure 3.18
Tsetse fly, Glossina *sp. (Photo courtesy of Peggy Greb, U.S. Department of Agriculture, Agricultural Research Service [USDA-ARS].)*

temperatures. Consequently, transfer between animals occurs more frequently in warmer months.

Being continuous obligatory ectoparasites, keds separated from a host generally die within 4 days due to lack of host warmth, moisture, and blood-meal source.

Tsetse Fly (Glossinidae), *Glossina* sp. (Figure 3.18)

Geographic Distribution: Sub-Saharan Africa

Veterinary Importance: Being restricted to western and central Africa, tsetse flies are of concern outside the geographical importance of what is covered in this book. However, they are included because of their importance in livestock production. Both females and males feed on blood. They are of most importance as vectors of the protozoan *Trypanosma brucci,* the causative agent of nagana, bovine trypanosomiasis. Affected animals become anemic and weak and stunted in growth. Infected animals are unsuitable for use as food. Early death may result from direct complications and/or secondary infection. Nagana continues to be a major deterrent to commercial livestock production in over one third of the African continent.

Hosts: Wide variety of vertebrate animals from reptiles to mammals

Description: Tsetse flies are narrow-bodied flies, 6 to 14 mm in length, and yellowish to dark-brown in color. They rest with their wings overlapping over the abdomen. Their forward-directing long proboscis can be easily seen when the flies are at rest.

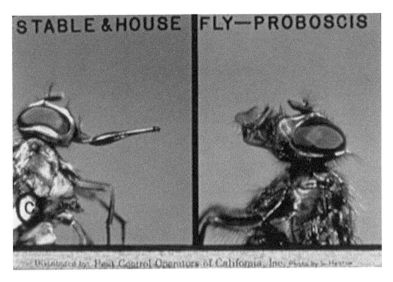

Figure 3.19
Fly mouthparts—sucking, sponging. (Photo reproduced with permission by Univar.)

Biology/Behavior: Similar to sheep keds, reproduction in tsetse flies is by larviposition. Newly emerged females mate within a day of emergence. Once mated, the female remains fertile for life. About 9 days after copulation, ovulation of a single egg occurs, and sperm are released to fertilize it. It hatches inside the female's oviduct within about 4 days. The larva passes through three growth stages (instars), feeding on secretions from modified accessory glands. At about 9 days, the fully developed third-instar larva is deposited on the ground where it pupates. Adults emerge from pupae after about 30 days. Females will deposit single larvae at 7- to 12-day intervals over their life span of 1 to 3 months.

NONBITING FLIES

Nonbiting flies, unlike biting flies, have soft, lobed mouthparts that are used to ingest liquid foods from surfaces (Figure 3.19). Although some species feed primarily away from animals, some actually obtain their nourishment from host secretions. Larval habitats are widely varied. The groups presented include only those that have some relationship to domestic animals.

Suborder Nematocera

Moth Flies (Psychodidae) (Figure 3.20)

Geographic Distribution: Worldwide

Figure 3.20
Moth fly, Psychodinae. (Photo reproduced with permission by Jim Kalisch, University of Nebraska, Lincoln.)

Veterinary Importance: The family Psychodidae has over 600 species worldwide. The subfamily Phlebotominae includes biting species known as sand flies, primarily in the genera *Lutzomyia* and *Phlebotomus*. Sand flies are known vectors of the protozoa that cause leishmaniasis and the virus causing vesicular stomatitis. Although leishmaniasis in humans is not of major concern in North America, some canine cases have been reported. Vesicular stomatitis is also not common in North America but has been detected in domestic animals and some wildlife species.

More common are the nonbiting moth flies in the subfamily Psychodinae. Most of the common species occur in the genera *Psychoda* and *Telmatoscopus*. Moth flies (also known as drain or filter flies) can be nuisance pests in confinement livestock housing and processing areas and in kennels when there are accumulations of moist/wet animal wastes and sewage/drain organic matter buildup. Adult moth flies tend to cling on walls, ceilings, support posts, and other areas and swarm when disturbed, causing an annoyance to people and animals.

Description: Adult moth flies (Psychodinae) are usually less than 5 mm long with dense hair on their body and wings. They are grayish to brown in color. The wings are large and ovate. They have long 12- to 16-segmented antennae. They have the general appearance of tiny moths. Larvae are grayish in color, elongate and legless, and up to 6 mm long. The head is well developed, and the last abdominal segment has a rigid siphon with a pair of spiracles at the tip.

Biology/Behavior: Moth flies breed in aquatic to semiaquatic habitats in wet organic matter associated with livestock waste pits, floor drains, and other areas where wet organic wastes accumulate. Adult females will deposit 20 to 100 eggs at a time. Larvae hatch in about 2 days. Larvae feed on moist organic debris. The lifecycle will take from 2 to 3 weeks to be completed. Adults may live for up to 2 weeks.

Suborder Brachycera (Infraorder Tabanomorpha)

Black Soldier Fly (Stratiomyidae) (Figure 3.21)

Species Name: *Hermetia illucens* (Linnaeus)

Geographic Distribution: Scattered worldwide and in the United States, more common in the southeastern states

Veterinary Importance: Black soldier fly larvae are most prevalent in poultry operations in the southeastern United States. If poultry manure is moist enough, soldier fly larvae can develop in large numbers. Dense, large populations can cause poultry manure to appear to "liquefy" and flatten, making it difficult for manure removal. Heavily infested manure often flows onto walkways. However, this does aid in the reduction of manure volume. Also, when these large populations develop, it drastically reduces the suitability of the manure for house fly development. Recent studies have shown that harvested black soldier fly larvae can be useful as a high-protein feed supplement and for use as live pet food (e.g., reptiles). Black soldier fly larvae are also found in association with carrion and have significant potential for use in forensic entomology.

Hosts: Immature black soldier fly larvae are found in animal wastes, sewage, and other decaying matter. They can be especially abundant in poultry manure from caged-egglayer houses.

Figure 3.21
Black soldier fly adult, Hermetia illucens.

Figure 3.22
Black soldier fly larva, Hermetia illucens. *(Photo reproduced with permission by Univar.)*

Description: Adult black soldier flies are large flies up to 15 mm long. They actually mimic in size, color, and appearance the organ pipe mud dauber wasp and its relatives. The antennae are elongate and wasp-like, and the fly's hind tarsi are pale. Also, they have two small translucent "windows" in the basal abdominal segments that resemble a narrow "wasp waist." Black soldier fly larvae reach up to 19 to 27 mm long and are somewhat flattened, roughly textured, and sandy to gray in color (Figure 3.22). Pupae resemble larvae but are immobile.

Biology/Behavior: Adult females prefer to oviposit their eggs in drier areas of manure or decaying organic matter. Adult females live for only 5 to 8 days and produce an average of 900 eggs. Eggs hatch in about 4 days, and the larvae develop through five instars over a 2-week period. The length of the larval developmental period is greatly extended in cooler temperatures and/or if food is lacking. Upon pupation, the pupal stage lasts 2 weeks or more. The entire lifecycle from egg to adult lasts from 35 to 60 days or more.

Suborder Brachycera (Infraorder Muscomorpha)

Face Fly (Muscidae) (Figure 3.23)

Species Name: *Musca autumnalis* (De Geer)

Geographic Distribution: Known from Europe south to Africa, Middle East to China, and in North America, primarily in the more temperate regions coast to coast

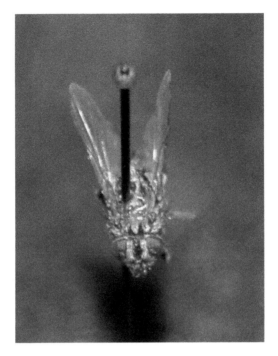

Figure 3.23
Face fly (female), Musca autumnalis.

Veterinary Importance: Face flies were first introduced into North America around 1951 and are now pests of range cattle in temperate regions of the United States and Canada, coast to coast. They will also readily feed on horses kept near cattle. Feeding activity of face flies on the face and head of host animals causes considerable direct annoyance (Figure 3.24). Host response varies, from eye twitching to head shaking to tail switching. Heavily infested animals tend to bunch together to avoid fly activity.

Although face flies have sponging mouthparts, they possess small rough spines on their mouthparts that can irritate host eye tissue. Infested animals often develop excessive eye secretions and redness. More importantly, face flies can serve as mechanical vectors of the bacterium *Moraxella bovis*, the causative agent of infectious bovine keratoconjunctivitis (IBK, pinkeye) in cattle. Face flies are also known vectors of spirurid nematodes (eyeworms) in the genus *Thelazia*. These nematodes, found in cattle and horses, are usually benign, causing no apparent clinical symptoms in their hosts. Another nematode, *Parafilaria bovicola*, is also vectored by face flies. This nematode causes subcutaneous lesions that result in tissue trimming loss at slaughter. This nematode is found in Sweden and South Africa and was detected in a cattle shipment to Canada. Thus far, however, it has not gotten established in North America.

Figure 3.24 (see color insert following page 132)
Face flies on cow.

Hosts: Cattle, bison, horses, and occasionally other animals

Description: Face flies closely resemble house flies in appearance. They are 6 to 10 mm long and have four black stripes on their thorax. They are slightly larger and more robust than house flies, and the underside of the abdomen is gray to black in appearance. House flies tend to be creamy yellow to white on the sides and underneath their abdomens. Eggs of face flies can be distinguished by having a brown-black respiratory stalk. Mature larvae are bright yellow (as compared to creamy white in color in other species), and the puparia are a distinctive white appearance, rather than brown, due to calcification.

Biology/Behavior: Face fly eggs are laid singly in fresh, undisturbed cattle or bison feces on pasture or rangeland prior to crust formation. Face flies do not lay eggs in aged manure piles around barns and stables, or in the disturbed, urine-saturated, and trampled droppings usually associated with beef feedlots and drylots. Eggs hatch in about 1 day. Larvae complete their development in 2½ to 4 days. As the larvae mature, they turn from white to yellow and move to the soil adjacent to the manure pat where they enter the pupal stage. In 5 to 7 days they will emerge as adults. The entire lifecycle is completed in about 2 weeks, depending on climatic conditions. Numerous generations per year may occur from April through October in temperate regions of North America.

Both adult male and female face flies feed on nutrients from plants and dung. Females require host proteins for egg development. They feed on facial secretions (e.g., tears, nasal mucous, saliva), blood from wounds, milk on calves' faces, vaginal discharge, placental fluid, and liquids from expelled placentas. Face flies do not constantly feed but rather have short feeding times and long periods of time away from their hosts. Only about 5% of a face fly population will be found on and around host animals at any given time, with 75% to 95% of flies observed on animals being females. Because of the extended times away from their hosts, and because face flies are strong fliers and are capable of traveling several kilometers a day, movement between widely separated herds may be frequent. This also contributes to horses often becoming infested when they are somewhat separated from cattle.

Newly emerged face flies at the end of the warm season will undergo a reproductive diapause, a state of physiologically controlled, cold-hardy quiescence, deterring reproduction until the following spring. They seek protective overwintering sites such as inside buildings and other structures. Clusters of overwintering face flies are often found in building lofts and attics. They can become domestic pests during this time by crawling on walls, windows, and floors during warm spells and when they become active just before leaving the hibernation site. Flies mate shortly after becoming active in the spring.

House Fly (Muscidae)

Species Name: *Musca domestica* (Linnaeus) (Figure 3.25)

Geographic Distribution: Cosmopolitan

Veterinary Importance: House flies are closely associated with humans, livestock, poultry, and companion animals. They are found in and around

Figure 3.25 (see color insert following page 132)
House fly, Musca domestica. *(Photo reproduced with permission by Jim Kalisch, University of Nebraska, Lincoln.)*

buildings and around organic wastes. Although adult house flies obtain nourishment from a variety of sources, they can feed directly on exposed blood, sweat, saliva, tears, and other body fluids of animals. In response to house fly annoyance, animals shake their heads, switch their tails, and avoid locations where house flies are abundant. Beyond annoyance, house flies do not cause any measureable harm to animal performance. On fly resting surfaces, house flies can degrade the appearance of facilities and animal products such as eggs by leaving unsightly regurgitation and fecal spots (fly specks).

House flies have significant potential for transmission of diseases of both veterinary and public health importance. House flies can serve as mechanical vectors of several disease agents because of their breeding and feeding habits. They can harbor disease agents both internally and externally on their body hairs, mouthparts, and feet. They serve as development hosts for two spirurid nematodes, *Habronema muscae* and *Draschia megastoma,* that cause gastric and cutaneous habronemiasis in horses. The house fly is also a host for the chicken tapeworm, *Choanotaenia infundibulum.* In humans, they have been implicated in the transmission of various pathogens causing enteric diseases (e.g., typhoid, paratyphoid, salmonellosis).

In poorly managed livestock and poultry facilities in which house fly populations can develop in high numbers, this may lead to public nuisance complaints when flies venture away from these production sites. This can be of significant economical consequences to livestock and poultry producers from legal actions from surrounding neighbor residences plagued with nuisance levels of house flies. There are cases on record of multi-million-dollar lawsuit awards from house fly nuisance associated with livestock and poultry operations where producers failed to adequately control house fly breeding in wastes produced in their operation.

Hosts: Very common around livestock, poultry, and companion animal facilities associated with accumulated wastes. It is also abundant around human wastes and garbage where adequate breeding conditions exist.

Description: Adult house flies range from 5 to 12 mm long and have a dull gray thorax and abdomen. The thorax has four longitudinal dark stripes on top. The abdomen of females has a checkered gray and black appearance dorsally and creamy yellow to white sides and underneath. House fly eggs are whitish in color, elongate, and about 1 mm long. There are three larval instars with mature larvae reaching from 10 to 15 mm long. House fly larvae and other muscoid larvae can usually be distinguished from each other by the arrangement of the posterior spiracular slits. House fly pupae are barrel shaped and red to brown in color.

Biology/Behavior: House fly females will deposit 50 to 150 eggs at a time in a cluster on a suitable breeding source at 2- to 4-day intervals. They are especially attracted to fecal sources but will lay eggs on other moist, decaying organic matter. As many as 500 eggs will be laid by a single female in her lifetime. Eggs will hatch in from 8 to 24 hours depending on temperature. Larval development takes from 3 to 7 days in warm weather. Upon maturity, prepupae are formed and migrate to dry areas surrounding their breeding source to pupate. The pupal stage lasts from 3 to 10 days, again dependent on temperature. Upon emergence, adult flies begin feeding within 24 hours and are capable of mating soon after. Most female house flies mate once, storing sperm in their spermathecae for multiple egg fertilization. Adults feed on a variety of organic matter from animal wastes and garbage to a large assortment of food sources. Solid food is liquefied from deposited fly saliva. It is then ingested with their sponging mouthparts.

The complete lifecycle takes from 7 to 14 days with two or more generations possible a month in warm weather. In temperate, cooler locations, house flies will overwinter as larvae and pupae protected in their breeding media where microbial activity prevents freezing. In warmer climates and heated livestock and poultry facilities, year-round active development will take place with all life stages of the house fly present.

Adult house fly behavior varies over the course of a day and in varying climatic conditions. During daylight hours, house flies favor areas near breeding and food sources. They frequent outdoor areas during warmer weather and often congregate indoors during cool, wet, or windy conditions. House fly movement is dependent on available breeding sources. Although house flies have been reported to fly several kilometers, most will remain within 1 to 1.5 km from their preferred breeding habitat where sufficient food and breeding media are available. At night, house flies can be found resting on various stationary objects from fences and vegetation outdoors to walls and ceilings indoors. During warmer weather, they often favor higher areas indoors as resting surfaces. Accumulation of "fly specks" is a telltale sign of where house flies congregate and rest in larger numbers.

Fannia sp. (Muscidae) (Figure 3.26)

Common Species:

> *F. canicularis* (Linnaeus)—little house fly
> *F. scalaris* (Fabricus)—latrine fly
> Others

Geographic Distribution: Cosmopolitan

Veterinary Importance: There are some 100 species of *Fannia* in North America, with the little house fly and latrine fly being most prevalent. Although they are rarely seen feeding directly on animals, they can be attracted to animal sweat, mucous, and fecal material adhering to the skin. They are more of an annoyance to people and animals, especially in poultry facilities and confined cattle and dairy operations, where they are often more abundant.

Hosts: Common around confined livestock and poultry operations

Figure 3.26
Fannia *sp. adult fly.*

Description: *Fannia* sp. generally resemble house flies but are smaller (4 to 8 mm long). They have a dark thorax and abdomen with varied yellow markings. Larvae and pupae are quite distinctive, having characteristic lateral and dorsal projections (Figure 3.27).

Biology/Behavior: *Fannia* breed in a wide variety of decomposing organic matter from animal, chicken, and human wastes. Being susceptible to desiccation, they are more abundant in semiliquid situations. Eggs are deposited on a suitable wet, decaying breeding surface where the larvae develop. They pupate in drier areas surrounding the breeding source. The complete lifecycle takes from 15 to 30 days. Adults tend to be more abundant in the cooler months of spring and fall. Adults can often be easily detected by their characteristic hovering behavior in flight. In cooler climates, they generally overwinter as larvae and as adults in warmer regions.

False Stable Fly (Muscidae) (Figure 3.28)

Species Name: *Muscina stabulans* (Fallen)

Geographic Distribution: Worldwide

Veterinary Importance: The false stable fly can be found breeding around animal wastes. Adults have similar feeding habits as house flies and present similar nuisance to animals and people and pose a concern in the mechanical transmission of food-borne pathogens. Larger populations may also lead to annoyance to residents near livestock operations and serve as grounds for legal action if adequate control procedures are not taken.

Figure 3.27
Fannia *sp. larva. (Photo reproduced with permission by Univar.)*

Figure 3.28
False stable fly, Muscina stabulans. *(Photo reproduced with permission by Patrick Jones, Purdue University.)*

Hosts: Can be found associated with various animal wastes

Description: False stable flies are somewhat robust flies, 8 to 12 mm long, with brownish-black bodies. The tip of the scutellum is reddish-orange. Mature larvae reach 13 mm long. Pupae resemble that of house flies, being barrel shaped and reddish-brown in color.

Biology/Behavior: Habits of false stable flies are similar to house flies. Eggs are laid in accumulated, moist animal manure or other decaying organic matter. Young larvae are primarily scavengers, and older larvae become predaceous, feeding on other fly larvae. The entire lifecycle takes from 4 to 6 weeks. In warmer climates, all life stages may be found throughout the year. In colder climates, they overwinter primarily as larvae and pupae under breeding sources protected from freezing. Adults have also been found overwintering in protected areas in a prereproductive diapause.

Garbage/Dump Flies (Muscidae) (Figure 3.29)

Species Names:

Hydrotaea leucostoma (Wiedamann)—black garbage fly
H. aenescens (Wiedamann)—black dump fly

Geographic Distribution: Genus is widely distributed worldwide

Veterinary Importance: These flies are generally not considered of significant veterinary importance. They can be of annoyance if abundant, primarily outdoors. In some confinement livestock and poultry operations, they can be considered beneficial because *Hydrotaea* larvae are predators of other flies, including house flies.

Figure 3.29
Black garbage fly, Hydrotea leucostoma.

Hosts: Can be common around confined livestock and poultry facilities and refuse dumps.

Description: These flies are from 4 to 7 mm long with distinctive shiny black coloration. Larvae and pupae are similar in appearance to those of house flies but are a little smaller.

Biology/Behavior: *Hydrotaea* sp. breed in similar habitats as house flies, usually preferring somewhat wetter conditions. Larvae feed as scavengers and will prey on other fly larvae. Because of this, they have been used in the biological control of house flies, especially in caged-egglayer operations. A complete lifecycle takes from 10 to 14 days under optimal conditions. After emergence, these flies prefer to stay in dark locations close to the ground. In confined livestock and poultry facilities, they often stay right at or near the manure surface. They rarely become a nuisance pest as do house flies because they have a tendency for short, low-height flights within their preferred breeding habitats.

Eye Gnats (Chloropidae) (Figure 3.30)

Species Name: *Hippelates* sp.

Distribution: Numerous species worldwide

Figure 3.30
Hippelates *sp.*

Veterinary Importance: *Hippelates* sp. gnats are bothersome pests of people, livestock, and companion animals. They can be of great annoyance to dogs. They often swarm around their hosts and readily feed on fluids around eyes, genitals, sores, and wounds. They have sponging mouthparts that have tiny spines that can irritate tissue around the eyes.

Hosts: Wide variety of mammals

Description: *Hippelates* sp. are very small flies about 2 mm long. They are from brownish to shiny black in appearance. They are distinguished from other gnats by having a large, black, curved spur on the hind tibia. The antennae are short. Larvae are white and maggot-like, less than 3 mm long. The brown pupae are cylindrical with depressed ends.

Biology/Behavior: Eye gnats are common in rural areas, meadows, fields, and other places where there are considerable areas of grass. Larvae primarily feed on roots of grass and other plants, decaying vegetation, and excrement. Eggs laid by females hatch in about 3 days, and the larval stage lasts about 2 weeks. The lifecycle takes 2 to 3 weeks or longer, depending on soil temperature and abundance of food. Adults are active from May to October in temperate climates and from March to November in warmer climates.

Rat-Tailed Maggot or Drone Fly (Syrphidae) (Figure 3.31)

Species Name: *Eristalis tenax* (Linnaeus)

Distribution: Cosmopolitan

Figure 3.31
Rat-tailed maggot adult, Eristalis tenax.

Figure 3.32
Rat-tailed maggot larva, Eristalis tenax.

Veterinary Importance: Adult drone flies are not of veterinary importance. They feed on flower pollen and nectar and do not bother animals. The larvae, known as rat-tailed maggots, are often found in manure-polluted water and around confined livestock operation waste lagoons. They can also be found in human wastewater treatment facilities.

Hosts: Larvae can be found in animal waste lagoons, wet manure, and sewage treatment plants.

Description: Adult drone flies are covered with fine yellow hairs and resemble honeybees, except they are true flies and have one pair of wings. They cannot bite or sting. The larvae are called rat-tailed maggots because of their long, retractable caudal segment that possesses the posterior spiracles. This segment extends two to three times the length of the rest of the larval body (Figure 3.32). This air tube allows the aquatic larvae to breathe air from the surface while the larvae are emerged in their aquatic habitat.

Biology/Behavior: Eggs are laid directly in the wet, rich organic matter habitat of waste lagoons, sewage treatment ponds, and other similar aquatic polluted sites. Larval development can take several months to complete. Maggots will migrate to drier sites surrounding the larval habitat to pupate.

———————————————— *CHAPTER 4*

MYIASIS

TABLE OF CONTENTS

Myiasis is the invasion of organs and/or tissues of living vertebrates by the larval stages of flies (Diptera). It encompasses the feeding on the host's living or dead, necrotic tissues. A diversity of fly species is involved in myiasis. As a form of parasitism, myiasis can cause significant harm to animals and can lead to death and in some cases to less severe effects leaving little or no tissue damage. Overall, however, myiasis should be considered of significant importance to livestock and companion animals.

CLASSIFICATION

Myiasis can be classified clinically according to tissue invaded by the fly larvae or in terms of the host–parasite relationship. In cutaneous myiasis, infestation of fly larvae of dermal or subdermal tissue of the skin is involved. When open sores or wounds are infested, this is referred to as traumatic myiasis. Gastrointestinal myiasis is invasion of the host's alimentary tract. Enteric myiasis is invasion specifically in the host's intestinal tract. Urogenital myiasis is invasion of the urethra/genitalia. Ocular or ophthalmic myiasis is invasion of eye tissue. Nasopharyngeal myiasis is invasion of nasal passages and sinuses and/or pharyngeal cavities. Auricular myiasis is the invasion of ears. In furuncular myiasis, fly larvae form boils or burrows in their host.

In classifying myiasis on the host–parasite relationship of the fly to its host, three types of myiasis are generally recognized: accidental, facultative, and obligatory. Accidental myiasis generally occurs when fly eggs or larvae contaminate food ingested by a host. In most cases, the fly species are not parasitic and pass through the host's alimentary tract. In facultative myiasis (sometimes called secondary myiasis), the fly species involved can develop in both living and dead hosts. In living hosts, the fly larvae typically feed upon dead, necrotic tissue around wounds and open sores. Typical of this type of myiasis are blow flies that normally develop in carrion. In obligatory myiasis (sometimes called primary myiasis), the fly species involved depend on a living host to complete their development and usually feed directly on living tissue.

FLIES INVOLVED IN OBLIGATORY MYIASIS

Screwworm (Calliphoridae) (Figure 4.1)

Species Name: *Cochliomyia hominivorax* (Coquerel)

Geographic Distribution: Found originally throughout tropical and subtropical areas of the Americas from the southern United States to Chile. With the success of an eradication program, this species has been eliminated from the United States, Mexico, and most of Central America.

Veterinary Importance: *Cochliomyia hominivorax*, considered as the New World screwworm, has been known from North America from ancient times. It is a major pest of domestic animals, especially cattle and other livestock. Screwworm larvae feed directly on living tissue of their host. Infestations generally start on open wounds on warm-blooded animals. Infested wounds range from deep scratches, arthropod bites, and lacerations to results of livestock husbandry practices such as castration, dehorning, branding, and shearing wounds. In the southwestern United States, the Gulf Coast tick, *Amblyomma maculatum*, which feeds primarily on the ears of cattle causing

Figure 4.1
Primary screwworm, Cochliomyia hominovorax. *(Photo reproduced with permission by Marcelo de Campos Pereira, University of Sao Pãulo, Brazil.)*

significant lesions, is a common site of screwworm strike. Untreated navels of newborn calves can also be infested. If infestations are not treated, screwworm cases can lead to death from tissue and vital organ destruction and secondary infections. Screwworm maggots will literally eat the host alive. As the feeding damage progresses, this attracts more gravid female screwworm flies to deposit more eggs, causing a self-perpetrating infestation (Figure 4.2).

The severity of this situation to the U.S. cattle industry in the 1900s led the U.S. Department of Agriculture (USDA) to focus efforts in eradicating the

Figure 4.2
Screwworm damage on cow. (Photo courtesy of U.S. Department of Agriculture, Agricultural Research Service [USDA-ARS].)

screwworm from the United States and further south. Millions of dollars were being lost annually due to treatment costs and animal loss. In 1962, when the eradication program was initiated in the southwestern United States, a major screwworm outbreak resulted in 49,000 reported cases in livestock in Texas alone. With additional outbreaks in 1968, 1972, and 1976, it was estimated that there were a total of nearly 2 million cases with losses of $300 million to the Texas economy.

In the 1950s, Edward F. Knipling and Raymond C. Bushland, working with the USDA, developed a sterile screwworm technique for controlling this pest. It centers on a unique reproductive handicap in which female *C. hominivorax* flies mate only once in their life. Focusing on clinically sterilizing huge numbers of males as breeding season approached and releasing them in mass to outcompete with fertile males, this could lead to the majority of female flies laying sterile eggs. A method was successfully developed by irradiating mass reared males and releasing them by aircraft. Tested in Florida, this resulted in the successful eradication of the screwworm by 1959. Thereafter, the program was expanded throughout the southern United States in the geographic range of the screwworm. Successful eradication in the United States was achieved in 1966. After the additional outbreaks in 1968, 1972, and 1976, the program has maintained complete eradication since the mid-1980s.

In 1972, Mexico and the U.S. Department of Agriculture, Animal and Plant Health Inspection Service (USDA-APHIS) initiated a joint screwworm eradication program to establish a biological barrier farther south, the Isthmus of Tehuantepec in southern Mexico. The program has since expanded with the goal of covering the entire Central American Isthmus and Panama and eventually reaching the Darien Gap separating Panama (Central America) and Columbia (South America).

In conjunction with the aerial release of sterile males are surveillance efforts made for detecting active infestations, as well as modifying management practices to reduce potential infestations. These range from calving, branding, dehorning, and other practices during the winter (fly-free) months, and control efforts such as routine treatment of wounds with insecticides, to using insecticide-impregnated ear tags, to controlling ticks that could cause fly strike. A screwworm bait product, Swormlure, was also developed and dispersed for controlling adult screwworm populations in infested areas.

Hosts: Nearly all warm-blooded animals, particularly mammals, including humans, cattle, horses, sheep, pigs, dogs, and others

Description: The adult screwworm fly is metallic-bluish in color with a yellowish orange face and three dark stripes on the thorax. The base of the wing stem vein (radius) bears a row of bristle-like hairs on the dorsal side. Larvae appear pinkish in color and have prominent rows of spines around the body. The tracheal trunks leading from the posterior spiracles have a distinctive dark pigmentation extending forward for a few segments. Mature third instar larvae reach a length of 15 to 17 mm.

Biology/Behavior: Gravid adult female screwworm flies are attracted to fresh wounds where they oviposit 100 to 300 eggs in a cluster on the edge of the wound at 2- to 3-day intervals. They produce up to 3000 eggs during their lifetime. Females commonly feed at the wound to obtain protein for producing their next egg mass. After a 12- to 24-hour incubation period, the eggs hatch, and the larvae start to feed gregariously, burrowing head-down into the living tissue. As the wound progresses, other female screwworm flies are attracted to lay more eggs. The larvae develop rapidly to maturity in 4 to 12 days and then drop to the ground to pupate. The pupal stage lasts from about 1 week during the summer to 3 months during the winter. Upon emergence, adult flies mate (females typically only mating once in their lifetime), and mated females lay their first batch of eggs 5 to 10 days postemergence. The entire lifecycle takes around 24 days during the summer and may be extended under cooler conditions. There is no true diapausing stage. Screwworms cannot survive over the winter in cool temperate climates. In the United States, the northernmost overwintering range is about at the latitude of San Antonio, Texas.

Cattle Grubs (Family Oestridae, Subfamily Hypodermatinae)

Species Names:

> *Hypoderma lineatum* (de Villers)—common cattle grub (Figure 4.3)
> *Hypoderma bovis* (Linnaeus)—northern cattle grub (Figure 4.4)

Geographic Distribution: Holarctic, widespread throughout the northern hemisphere. In North America, the common cattle grub is prevalent in the United States and Canada (except southern Texas and Alaska). The northern cattle grub is found primarily in the northern United States and Canada.

Veterinary Importance: Cattle grubs have been of major economic importance to the U.S. cattle industry. In 1976, annual losses in the United States were estimated to be $360 million. Losses result from damaged hides, reduced weight gains, lowered weaning weights, tissue/meat damage from trim loss and carcass downgrading at slaughter, and self-injury by panicked cattle running to escape ovipositing adult female flies. This panic behavior is known as gadding. In addition to this erratic running behavior, cattle will stand in ponds and streams and seek dense vegetation to avoid the flies. Common names given to these flies are "bomb" flies for northern cattle grubs, and "heel" flies for common cattle grubs reflective of adult fly disturbance of cattle. Improper timing of chemical control treatment can also have detrimental effects on cattle. If cattle grubs are killed with systemic insecticide treatment after they have migrated to the gullet or spinal column of the host animal, decomposing

Figure 4.3 (see color insert following page 132)
Common cattle grub adult, Hypoderma lineatum.

Figure 4.4
Northern cattle grub adult, Hypoderma bovis.

remains of cattle grub larvae may produce tissue swelling, resulting in suffocation or paralysis.

With the introduction of certain systemic parasiticides, especially the avermectins, cattle grub populations have declined significantly in the United States and Canada in recent years.

Hosts: Cattle

Description: Adult female cattle grubs are 13 to 15 mm long and bee-like in appearance. They are covered with dense hair with a distinctive light–dark color pattern. In *H. bovis*, the thorax is more hairy than that of *H. lineatum*. Mouthparts are small and nonfunctional.

Cattle grub larvae have three growth stages with younger larvae almost white, changing to yellow and light brown, to almost black when mature. When mature, they are robust and up to 25 mm long. They have distinctive flat tubercles and small spines present on all but the last body segment.

Biology/Behavior: Adults of the common cattle grub are active from February (in warmer regions) through June/July. Adults of the northern cattle grub usually appear about a month later and may be active up to early September in regions where they occur. Upon emergence from pupae, male flies aggregate and await females. The males will meet females in flight, and both quickly fall to the ground to complete mating. After mating, females begin searching for a host to lay eggs. Adults usually live only for around 2 to 3 days.

Egg-laying behavior differs between the two species. Female common cattle grubs typically attach eggs in rows of 3 to 10 or more on a single hair shaft of the lower leg or other lower-body region of the host. Females of the northern cattle grub dart in and out of the lower region of cattle to deposit several eggs, one at a time. This egg-laying activity, especially by the northern cattle grub, is what causes the gadding behavior in cattle. The total number of eggs deposited by individual cattle grub females ranges from 500 to 800.

Eggs of both cattle grub species hatch within 3 to 7 days. Newly hatched larvae migrate down the hair shaft to the skin and penetrate directly through the skin or through a hair follicle. The larvae then spend from 4 to 6 months migrating through connective tissue of the host. They secrete digestive enzymes that aid in their migration. In the first 1 to 2 months, they are still quite small (about the size of a grain of rice). At about this time, common cattle grubs congregate in the connective tissue of the esophagus, and northern cattle grubs generally congregate in tissue surrounding the spinal column. At these sites, the larvae feed and increase in size up to six times before continuing their migration to the back of their host. These first-stage larvae upon reaching the back create a small hole in the skin. Shortly after the grubs arrive at the back, the host responds by surrounding each grub in a cyst called a "warble." The warble isolates the grub within the host and provides the grub a rich source of nutrient from host body secretions. In this warble, the grub develops through

Figure 4.5
Cattle grub in back.

two subsequent instars increasing in size up to 15 mm in length when mature (Figure 4.5). Upon maturity, the grub exits the warble through the breathing hole and fall to the ground to pupate. The pupal stage lasts from 1 to 3 months. There is only one generation per year.

The first grubs usually appear at the back 32 to 34 weeks after eggs are laid and continue to arrive for 6 more weeks. Common cattle grubs arrive first from September to January (depending on location) with northern cattle grubs appearing 4 to 6 weeks later. Natural larval mortality is quite high with only up to 5% of the eggs laid giving rise to adults. In untreated cattle, average numbers of cattle grub larvae range from 5 to 20 per infested animal in locales where cattle grubs occur. However, up to 250 grubs per animal have been recorded.

Given the lifecycle and migration patterns of cattle grubs, timing of control using systemic insecticide treatments is critical. Cattle grubs should be killed before they reach the esophagus and/or spinal column areas to avoid host swelling and possible suffocation or paralysis. Geographical "cutoff" dates are established in different regions of the country where cattle are located or shipped from. These dates can be obtained from individual State Cooperative Extension Service specialists or publications.

Horse Bot Flies (Family Oestridae, Subfamily Gasterophilinae) (Figure 4.6)

Species Names:

Gasterophilus intestinalis (De Geer)—horse bot fly

Figure 4.6
Adult horse bot fly, Gasterophilus intestinalis.

> *G. nasalis* (Linnaeus)—throat bot fly
> *G. haemorrhoidalis* (Linnaeus)—nose bot fly

Geographic Distribution: Worldwide

Veterinary Importance: Of the *Gasterophilus* species of horse bots, these three species are the most common and are found worldwide as parasites of equids (horses, mules, donkeys). Infestations are very common among equids, at least at some part of their life. Light infestations often have little detrimental effect on the host. However, bots may cause significant concern and harm to their host in a number of ways. Oviposition behavior of adult female bot flies can cause noticeable disturbance and panic of the host animal. Horses often will react violently to ovipositing females and injure themselves or others in avoidance behavior. Following oral entry, first-instar bot larvae burrowing in oral tissue can cause irritation, pus pockets, loose teeth, and appetite loss. When larvae pass in the esophagus and stomach and attach to the inner surface of the gastrointestinal mucosa and stomach wall, this can cause inflammation, sloughing of tissue, ulcerations, modular mucosal and stomach proliferation, abscesses, peritonitis, and interference of digestion (Figure 4.7). In heavy infestations, general debilitation and rectal prolapse may result. Infested animals may experience reduced weight gains and often become more susceptible to other infections. In extreme infestations, death has been known to occur.

Hosts: Equids (horses, donkeys, mules)

Description: Adult *Gasterophilus* bot flies superficially resemble honeybees in size and coloration. They range in size from 11 to 15 mm long with the body densely covered with yellowish hairs. The female ovipositor is

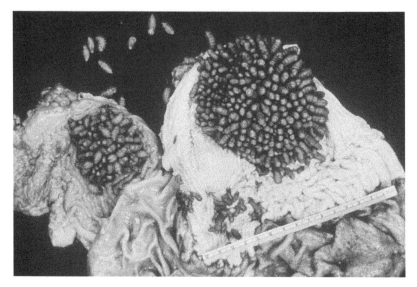

Figure 4.7
Horse bot fly larvae in stomach of horse. (Photo courtesy of University of California.)

distinct and protuberant. Adults have nonfunctional mouthparts and do not bite or feed. Larvae are grub-like in appearance with mature larvae reaching a length up to 20 mm and yellowish in color. They are equipped with mouth hooks for attaching to their host and have rows of heavy spines at the front of each segment.

Biology/Behavior: Adult *Gasterophilus* flies are short lived with an effective life span of only a few days. The female adults mate and begin oviposition soon after emergence. Eggs are attached to the host hair shafts. Depending on species, from 150 to 1000 eggs can be deposited by a single female all within a few hours if hosts are available and weather permits. Females will hover around the host animal and dart in to deposit one egg at a time. Species-specific biology is as follows.

Horse Bot Fly: The horse bot fly is the predominant species in North America. The yellowish-gray eggs are usually attached on the inner forelegs but can also be found on the outside of the legs, mane, and flanks. From 500 to 1000 eggs can be deposited by a single female. Incubation of eggs is around 1 to 2 weeks, after which they are stimulated to hatch from the warmth, moisture, and friction from the host's tongue as they groom themselves. The newly hatched larvae burrow into the dorsal mucosa of the host tongue. After several days to a few weeks, they molt into the second stage and move to the mucosa of the esophageal portion of the stomach where they attach to feed on tissue exudates. They molt into the third stage and remain attached for 8 to 10 months. Mature larvae will then detach and pass out with the feces. Pupation

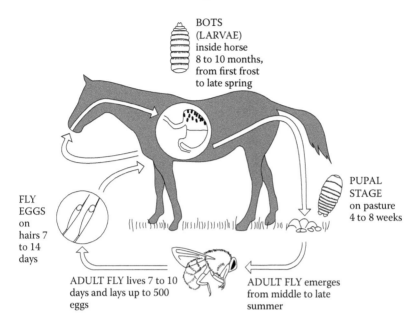

BOTS
(LARVAE)
inside horse
8 to 10 months,
from first frost
to late spring

PUPAL
STAGE
on pasture
4 to 8 weeks

FLY
EGGS
on
hairs 7
to 14
days

ADULT FLY lives 7 to 10
days and lays up to 500
eggs

ADULT FLY emerges
from middle to late
summer

Figure 4.8
Horse bot fly lifecycle. (Illustration courtesy of Sarah Cox, Purdue University.)

occurs in the soil or dry dung soon after larvae drop from the host. The pupal
stage lasts about 4 to 8 weeks. Adult flies emerge, mate, and quickly resume
activity near equid hosts (Figure 4.8).

Throat Bot Fly: Throat bot flies attach their whitish-yellow eggs in small
batches underneath the jaw of their hosts. They produce from 400 to 500
eggs. Eggs hatch in 4 to 5 days, and newly hatched larvae crawl into the mouth
and penetrate soft tissue. After about 3 weeks they molt into the second stage
and migrate from the mouth and attach to the pyloric region of the stomach or
anterior portion of duodenum. Here they grow and molt into the third stage.
Mature larvae pass out with the feces after about 10 to 11 months to pupate.

Nose Bot Fly: Nose bot flies attach their blackish stalked eggs to the hairs
on the upper and lower lips of horses. They produce an average of 150 to
200 eggs. Moisture from licking stimulates hatching of the larvae after about
2 days of incubation. First-stage larvae burrow into the epidermis of the lips
or in the tongue where they remain for about 6 weeks. After molting, the
second-stage larvae then move down the esophagus and attach to the pyloric
region of the stomach or duodenum where they molt into the third stage.
After several months, mature larvae move to the rectum and reattach close to
the anus. Then, after 2 to 3 days, they detach and are passed out in the feces
to pupate.

With all three species, there is usually only a single generation per year in
temperate regions.

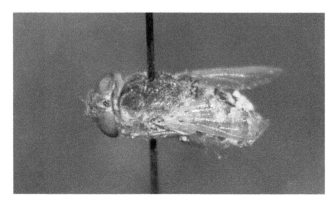

Figure 4.9
Sheep bot fly adult, Oestrus ovis.

Sheep Bot Fly (Family Oestridae, Subfamily Oestrinae) (Figure 4.9)

Species Name: *Oestrus ovis* (Linnaeus)

Geographic Distribution: Sheep-growing areas worldwide

Veterinary Importance: The sheep bot fly causes relatively low economic losses, but damage can be severe in individual animals. As they deposit larvae, adult sheep bot flies can annoy sheep, reducing grazing time and resulting in reduced weight gains and loss of condition. Annoyed sheep tend to bunch, shake their heads, stamp their feet, and rub their noses against the ground. Larval activity in nasal cavities causes symptoms ranging from mild discomfort, mucoid to bloody nasal discharge, sneezing, nose rubbing, and head shaking. Dead larvae in the sinuses can cause allergic and inflammatory reaction and sometimes bacterial infections. There is a low mortality associated with sheep bots.

Hosts: Sheep and goats

Description: Adult sheep bot flies are hairy and bee-like and yellowish-gray in color. They are about 10 to 12 mm long. Their mouthparts are reduced. Larvae are white when young and change to a slightly yellow or brown as they mature, with dark transverse bands on each segment. The ventral surface of each segment bears a row of small spines. When fully grown, they reach a length of 20 to 30 mm.

Biology/Behavior: Adult female sheep bot flies are oviparous—that is, they deposit newly hatched first-stage larvae, not eggs, that enter the nostrils of the host. They deposit up to 25 larvae at a time. Females are active during hot,

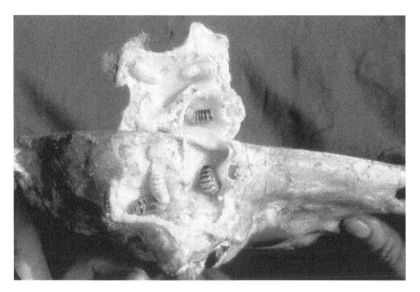

Figure 4.10
*Sheep bot fly (*Oestrus ovis*) larvae in sinuses of sheep.*

dry weather from late spring to autumn and may produce up to 500 larvae during their lifetimes. The small 1-mm first-stage larvae crawl up the nasal passages and attach to the mucous membranes using their mouth hooks and feed on mucous secretions. They then move to the frontal sinuses and molt into the second stage. Reaching the upper nasal passages and sinuses, they molt into the third stage to complete their development (Figure 4.10). Mature larvae reenter the nasal cavities where they are sneezed out by the host and drop to the ground where they pupate in the soil. The pupal stage will last from 10 to 70 days depending on temperature and geographic location. Larvae deposited in late spring to early summer may produce a second generation of adults in autumn. Larvae deposited later in summer tend to develop slower, dropping to the ground in late winter or early spring, and entering a diapause until late spring.

Flesh Fly (Sarcophagidae) (Figure 4.11)

Species Names

> *Wohlfahrtia vigil* (Walker)
> *W. magnifica* (Schiner)

Geographic Distribution: *Wohlfahrtia vigil* occurs in North America, primarily in Canada and the northern United States. *Wohlfahrtia magnifica*, an Old World species, occurs in the southern Paleartic region.

Figure 4.11
Adult flesh fly, Wohlfahrtia *sp.*

Veterinary Importance: *Wohlfahrtia vigil* has been known to parasitize mink and fox at fur ranches. They have also been found on rabbits, dogs, cats, and occasionally other animals and humans. Infestations are usually fatal to very young animals. Losses of kits (mink) as high as 30% have been observed at some ranches. May and June are the months of greatest concern for protection of young animals at these ranches. Forming boil-like lesions, *W. vigil* can cause a loss of appetite, fever, and restlessness of their host. As few as five larvae per animal can produce severe losses.

Hosts: Preferred hosts of *W. vigil* include mink, foxes, rabbits, and occasionally dogs, cats, ferrets, birds, and humans.

Description: Adults are large, 8 to 14 mm in length, grayish-colored flies with three dark stripes on the thorax and a number of distinct, separate, rounded dark patches on the abdomen. Larvae are maggot-like with third-stage larvae ranging from 7 to 16 mm long.

Biology/Behavior: As with other sarcophagid flesh flies, females are ovoviviparous depositing first-stage larvae. Larvae are deposited on the host near wounds or body orifices. The larvae quickly penetrate the skin forming boil-like lesions. The larvae feed and mature in 5 to 9 days, and then leave the wound and drop to the ground to pupate. The pupal stage lasts from 10 to 12 days. Up to 12 to 14 larvae can be found infesting a single host.

Figure 4.12
Rodent bot fly, Cuterebra buccata.

Rodent Bots (Family Oestridae, Subfamily Cuterebrinae) (Figure 4.12)

Species Names: *Cuterebra* sp.

Geographic Distribution: Western Hemisphere

Veterinary Importance: Species in the genus *Cuterebra* are primarily para-sites of rodents and rabbits but may occasionally infest dogs and cats. Damage caused by these parasites is usually not significant, but fatal cases due to infes-tations have been recorded in cats and dogs, especially in younger animals. In the warble formed by these bot larvae, a thin layer of necrotic tissue develops. When the larvae exit, the warble damage can occur to the skin or hide of the host animal. In wildlife hosts such as rabbits and squirrels, the edibility of meat is not impaired, but tissue adjacent to the warble usually needs to be trimmed away during cleaning and processing.

Hosts: Rodents and rabbits, and occasionally dogs and cats

Description: Adults are large, robust flies up to 20 mm long. They are cov-ered by dense, short hair and blue- to black-colored abdomens. They have reduced, nonfunctional mouthparts and do not feed as adults. Larvae are grub-like and have strongly curved mouth hooks and numerous strong body spines (Figure 4.13).

Figure 4.13
Cuterebra *larva. (Photo reproduced with permission by Jim Kalisch, University of Nebraska, Lincoln.)*

Biology/Behavior: Adult females lay eggs on the ground near the entrance of host nests or burrows, or near trails used by hosts. Upon egg hatch, larvae attach onto a host and enter the body directly through the skin or natural orifice. They will migrate subdermally in the host to seek a particular body region which tends to be host species specific, where they form pocket-like warbles. In rodents, the warble is often formed near the anus, scrotum, or tail, while some species prefer head, neck, or throat regions. In small rodents, such as in mice, the warbles can be as big as the rodent's head. Parasitism by more than one larva is common. In kittens and puppies, warbles are often seen in the head area (Figure 4.14). Larval development ranges from 3 to 7 weeks, depending on host species and climatic conditions. Mature larvae emerge from the skin and drop to the ground to pupate. In most species, there is only a single generation per year.

OTHER FLIES ASSOCIATED WITH OBLIGATORY MYIASIS

There are a few other species of flies involved with obligatory myiasis, but they have minimal importance as pests of domestic animals.

The reindeer or caribou warble fly, *Oedemagena tarandi* (Linnaeus), attacks reindeer, caribou, and musk ox. They are Holarctic in distribution in the northern latitudes. Their lifecycle is similar to cattle grubs in cattle in that eggs are deposited on lower portions of the host, and larvae penetrate the body and migrate to the host's rump.

Deer bot flies in the genus *Cephenemyia* are restricted to the Holarctic region and develop only in deer. Larvae develop in the pharyngeal and nasal

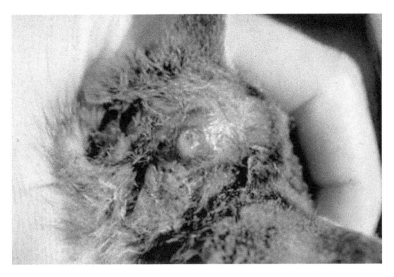

Figure 4.14
Kitten with Cuterebra *larva in head.*

passages of their host. Activity of larvae can cause nasal discharge, sneezing, coughing, and restlessness.

The human bot fly, *Dermatobia hominis* (Linnaeus), also known as the torsalo, is not found in the United States but is common in parts of Mexico and Central and South America. It is considered an important parasite of cattle in these regions. Larval stages can be found in several host species, including cattle, sheep, goats, pigs, buffalo, dog, cats, rabbits, and humans. It has a unique lifecycle. Adult flies catch various blood-sucking flies such as mosquitoes and attach their eggs to them. When the mosquito feeds on a warm-blooded host, the bot fly larvae hatch and penetrate the skin of the animal. They develop warbles in the subcutaneous tissue, and in 4 to 18 weeks when they mature, they leave the host to pupate on the ground. In cattle the cutaneous swellings can be pruritic, and breathing/exit holes may attract other myiasis flies, such as the screwworm fly. Infestations may result in hide damage and reduction in meat and milk production.

FLIES INVOLVED IN FACULTATIVE MYIASIS

Blow Flies (Calliphoridae) (Figure 4.15)

Species Names: Numerous species found in North America

> Green bottle flies—*Lucilia* sp.
> Blue bottle flies—*Calliphora* sp., others
> Black blow fly—*Phormia regina* (Meigen)

Figure 4.15
Calliphorid blow flies. (Photo reproduced with permission by Jim Kalisch, University of Nebraska, Lincoln.)

Secondary screwworm—*Cochliomyia macellaria* (Fabricius)
Hairy maggot blow fly—*Chrysomya rufifacies* (Macquart)
Others

Geographic Distribution: Blow flies are found throughout the world.

Veterinary Importance: Blow flies are carrion feeders and have important utility in forensic entomological investigations in assessing time intervals since death. On living animals, these flies are attracted to open sores and chronic nasopharyngeal or urogenital infections where maggots feed on dead tissue. They will also invade wounds caused by husbandry practices such as dehorning, castration, branding, shearing wounds, and other practices that may cause trauma to host tissue. Seepage from surgical procedures can also lead to infestation. Some species will be found in the same wounds fed upon by the primary screwworm and in feeding lesions caused by ticks.

In sheep, blow flies are attracted to and will lay eggs on fleece contaminated with feces, urine, blood, or body fluids. Common names given those species on sheep are fleeceworms and wool maggots (Figure 4.16). Different types of fly strike are observed when blow flies infest sheep. Breech strike occurs on wool around the crotch area that is soiled with urine, wet feces, or birth fluids. Body strike is often observed during wet seasons or when temperatures are hot and excessive sweating occurs. Head strike often occurs in rams involved in butting.

Figure 4.16
Wool maggot infestation.

Blow fly larvae and wool maggots, although they feed only on dead tissue, can cause significant harm to animals. Irritation of the skin caused by the larvae produces skin secretions and secondary bacterial infections. This can escalate the spread of the infestation over the host's body. It also attracts more blow fly females to lay eggs on the host. In sheep, the wool often falls away from severely infested areas. If not treated, blow fly and wool maggot infestations can dramatically decrease animal growth and productivity. Wool loss can be significant. In highly traumatized animals, death can occur.

Hosts: All warm-blooded animals are susceptible to infestation.

Description: Adult calliphorid blow flies range in size from 6 to 14 mm in length. Most species are metallic with coloration ranging from green to blue to bronze to black. Larvae are typical maggots and are usually white to cream in color. Mature larvae range from 8 to 23 mm long.

Biology/Behavior: Most species of blow flies have a similar lifecycle. Adult females are attracted to odors disseminating from a wound or soiled skin. Up to 100 to 200 eggs will be deposited by an individual female at a time directly on the wound or soiled substrate. Eggs hatch within a few hours, and the larvae feed and develop on the host for 3 to 5 days or longer, passing through three growth stages while feeding. Mature larvae will migrate from the feeding location and drop to the ground to pupate. Adults emerge after several days. The complete lifecycle takes from 5 to 10 days in warmer weather to 2 to 3 weeks or longer in cooler conditions. Different species will vary in their lifecycles.

Other Flies Causing Facultative Myiasis

Several other flies have occasionally been observed causing facultative myiasis in domestic animals. Flesh flies (Sarcophagidae), primarily in the genus *Sarcophaga*, although primarily associated with carrion or feces, have at times been found associated with wounds and accidental gastrointestinal myiasis. *Fannia* sp. (Muscidae) have been reported in cases of gastrointestinal and urogenital myiasis, and rarely in wounds. Among other flies, the black soldier fly (*Hermetia illucens*), cheese skipper flies (Piophilidae), house fly (*Musca domestica*), false stable fly (*Muscina stabulans*), and rat-tailed maggot (*Eristalis tenax*) have been observed in association with gastrointestinal and urogenital myiasis.

CHAPTER 5

LICE (ORDER PHTHIRAPTERA)

TABLE OF CONTENTS

Lice found on domestic animals are host-specific, permanent obligatory ectoparasites. They spend their entire life on their host. Lice are hemimetabolous, having egg, nymph, and adult stages. There are three nymphal stages. Lice are well adapted to spend their entire lifecycle on their host. Adult lice are small, about 0.5 to 8 mm long, dorsoventrally flattened, and wingless, and their legs and tarsi are adapted for maneuvering through fur, hair, and feathers. Adult female lice cement their eggs individually to the hair or feathers of their hosts.

The order Phthiraptera is broken down into two primary taxonomic groups: the Anoplura or sucking lice and the Mallaphaga or chewing or biting lice. Anoplura is classified as a suborder. Within the Mallaphaga there are three suborders: Amblycera, Ischnocera, and Rhynchophthirina. Rhynchophthirina is a small suborder including only a couple of African species and not discussed here. Amblycera and Ischnocera are discussed together as Mallaphaga.

In Table 5.1 are included the important lice of veterinary importance that are discussed in this chapter.

Adult Mallaphaga lice are adapted for chewing, possessing chewing mandibles. They are generally small lice (2 to 3 mm long) and have large heads, almost as wide as the rest of their body. These lice feed on a variety of skin fragments, hair, or feathers on their host. Some species will feed on blood exuding from scratches in the skin or from rasping action of their mouthparts. Both mammals and birds are fed upon by chewing lice.

Anoplura, the sucking lice, feed exclusively on mammals. They range in size from 0.5 mm to as large as 8 mm in length. The head is small, narrow, and elongated. Their mouthparts are modified for piercing the skin. The mouthparts consist of three stylets used to puncture the skin and a sucking tube for taking in the blood meal. At rest, the mouthparts are withdrawn into the head within a modified snout-like labrum. This snout is equipped with tiny hook-like teeth to attach to the host while feeding.

In general, there are seasonal population trends of both chewing and sucking lice. Typically larger infestations occur during cooler seasons when hosts are prone to cool weather stress, have thicker hair coats, undergo molting, and alter hormone levels. Lice typically are transferred from animal to animal by direct contact. Transfer of lice from an infested mother to her offspring, nest or pen sharing, and during mating are common ways that lice move from one host to another. Some species of lice can disperse by phoresy, hitching a ride on other arthropods such as blood-feeding flies from one host to another.

Host specificity ranges from a single host species, such as with the hog louse, to some lice species parasitizing two or more closely related species. Dog lice, as an example, may parasitize other canines such as foxes, wolves, and coyotes, and the horse sucking louse parasitizes horses, donkeys, and mules.

CATTLE LICE

Cattle Biting Louse (Mallaphaga, Suborder Ischnocera; Family Trichodectidae) (Figure 5.1)

Species Name: *Bovicola bovis* (Linnaeus)

Geographic Distribution: Worldwide

Veterinary Importance: The cattle biting louse is the only chewing louse found on cattle in the United States. On cattle, they are often found at the base of the tail, back line, and shoulders. As louse numbers increase, populations often spread to other parts of the body. They are often seen in patches of several lice together where extensive feeding occurs. They feed primarily on shed skin scales, but chewing activity causes lesions on the skin, similar in appearance to mange caused by mites. Infested animals may appear shaggy,

Table 5.1
Lice of Veterinary Importance (Order: Phthiraptera)

Species	Common Name	Host(s)
Mallaphaga—Chewing Lice		
Suborder Amblycera		
Family Boopiidae		
Heterodoxus spiniger (Enderlein)	Dog biting louse	Canines
Family Gyropidae		Guinea pigs
Family Menoponidae		
Menopon gallinae (Linnaeus)	Shaft louse	Domestic fowl
Mencanthus stramineus (Nitzsch)	Chicken body louse	Domestic fowl
Suborder Ischnocera		
Family Philopteridae		
Chelopistes meleagridis (Linnaeus)	Large turkey louse	Turkeys
Cuclotogaster heterographus (Nitzsch)	Chicken head louse	Domestic fowl
Gonicocotes gallinae (De Geer)	Fluff louse	Domestic fowl
Goniodes gigas (Taschenberg)	Large chicken louse	Domestic fowl
Lipeurus caponis (Linnaeus)	Wing louse	Domestic fowl
Family Trichodectidae		
Bovicola* bovis (Linnaeus)	Cattle biting louse	Cattle
B. caprae (Gurlt)	Goat biting louse	Goats, sheep
B. equi (Denny)	Horse biting louse	Equines
B. limbatus (Gervais)	Angora goat biting louse	Angora goats
B. ovis (Schrank)	Sheep biting louse	Sheep
Felicola subrostrata (Burmeister)	Cat biting louse	Cats
Trichodectes canis (De Geer)	Dog biting louse	Canines
Anoplura—Sucking Lice		
Family Haematopinidae		
Haematopinus asini (Linnaeus)	Horse sucking louse	Equines
H. eurysternus (Nitzsch)	Short-nosed cattle louse	Cattle
H. quadripertusus (Fahrenholz)	Cattle tail louse	Cattle
H. suis (Linnaeus)	Hog louse	Swine
Family Linognathidae		
Linognathus africanus (Kellogg and Paine)	African blue louse	Sheep, goats
L. ovillus (Neumann)	Face and body louse	Sheep
L. pedalis (Osborn)	Sheep foot louse	Sheep
L. setosus (Olfers)	Dog sucking louse	Canines

Table 5.1 (Continued)
Lice of Veterinary Importance (Order: Phthiraptera)

Species	Common Name	Host(s)
L. stenopsis (Burmeister)	Goat sucking louse	Goats
L. vituli (Linnaeus)	Long-nosed cattle louse	Cattle
Solenopotes capillatus (Enderlein)	Little blue cattle louse	Cattle

* *The genus* Bovicola *is referred to as* Damilinia *in some taxonomic references.*

Figure 5.1
Cattle biting louse, Bovicola bovis. *(Photo reproduced with permission by Jim Kalisch, University of Nebraska, Lincoln.)*

discolored, and ragged in heavily infested areas of the skin. Animals spend noticeable time licking and rubbing themselves.

Hosts: Specific on cattle

Description: The cattle biting louse is reddish-brown in color with a broad triangular, flat head, and dark transverse bands on the lighter-colored abdomen.

Biology/Behavior: Eggs are glued to the base of hairs of cattle. They typically hatch in 7 to 10 days. Each of the three nymphal instars develops for 5 to 6 days for a total development time of 15 to 20 days. Total development from egg to adult emergence takes about 1 month. Adults can live as long as 10 weeks.

Short-Nosed Cattle Louse (Suborder Anoplura; Family Haemotopinidae) (Figure 5.2)

Species name: *Haematopinus eurysternus* (Nitzsch)

Figure 5.2
Short-nosed cattle louse, Haematopinus eurysternus.

Geographic Distribution: Temperate regions worldwide

Veterinary Importance: Infestations of the short-nosed cattle louse are most frequently found on mature animals, reaching peak population levels in winter. In winter they are often seen feeding along the top of the neck, dewlap, brisket, base of the horns, and base of the tail. In higher infestations, they may cause anemia and loss of condition.

Hosts: Cattle

Description: The short-nosed cattle louse is the largest louse found on North American cattle. Adults range from 3.5 to 4.5 mm in length and 1.5 mm wide. They are easily recognized by their gray-black body and relatively blunt head. There are noticeable conical bumps along the sides of the body.

Biology/Behavior: Adult females attach one to four opaque brown eggs to hairs on cattle per day for about 2 weeks. After an incubation period of 9 to 19 days, nymphs emerge. Nymphs develop through three instars reaching adulthood in about 2 weeks. The average lifecycle from egg to adult takes from 20 to 40 days. Adults live for about 10 to 15 days.

Cattle Tail Louse (Suborder Anoplura; Family Haematopinidae) (Figure 5.3)

Species name: *Haematopinus quadripertusus* (Fahrenholz)

Geographic Distribution: Warmer regions worldwide; common in the United States, especially in the Gulf Coast states.

Figure 5.3
Cattle tail louse, Haematopinus quadripertusus. *(Photo reproduced with permission by Marcelo de Campos Pereira, University of São Paulo, Brazil.)*

Veterinary Importance: Populations of the cattle tail louse are found primarily in the long hair around the tail. In heavy infestations, tail hairs become matted with lice eggs, and tail heads may be shed of hair. Extensive infestations may cause anemia in cattle. Cattle show poor condition, slower weight gain, low vitality, and reduced milk production. In Florida, this louse is considered the most damaging louse species on cattle.

Hosts: Primarily cattle, sometimes found on zoo animals. Infestations are more common on zebu cattle, *Bos inducus,* than on *Bos taurus* cattle.

Description: The cattle tail louse resembles the short-nosed cattle louse but is larger. The head, thorax, and abdomen are dark brown.

Biology/Behavior: Adult females oviposit eggs on the tail hairs. Large numbers of eggs are often observed. Eggs incubate from 9 to 25 days, depending on the season (longer in cooler weather). Nymphs will move over the body surface and can be found in the area of the eyes, muzzle, and vulva. Third-instar nymphs migrate to the tail head region. Adults are typically confined to the tail head. The complete lifecycle takes as few as 25 days to complete. This species is most abundant during the summer, unlike other species that are more common during the cooler months.

Long-Nosed Cattle Louse (Suborder Anolplura; Family Linognathidae) (Figure 5.4)

Species Name: *Linognathus vituli* (Linnaeus)

Geographic Distribution: Worldwide

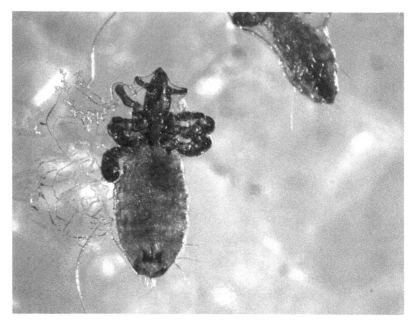

Figure 5.4
Long-nosed cattle louse, Linognathus vituli.

Veterinary Importance: The long-nosed cattle louse is most common on calves and dairy cattle and rarely occurs in large numbers on mature animals. Preferred infestation sites are the dewlap and shoulders. As weather warms in the spring, populations are often confined to the shoulder areas. This louse can be of economic concern, especially on younger animals where it can affect weight gain efficiency.

Hosts: Cattle

Description: The long-nosed cattle louse has a slender, pointed head with a bluish-black narrow body. Adults are around 2.5 mm long.

Biology/Behavior: Adult females lay a single egg per day. Eggs hatch in 10 to 15 days. Nymphs develop through three growth stages over a 2-week period. The entire lifecycle takes 20 to 30 days to complete.

Little Blue Louse (Suborder Anoplura; Family Linognathidae) (Figure 5.5)

Species Name: *Solenoptes capillatus* (Enderlein)

Geographic Distribution: Worldwide; restricted to areas with domestic animals

Lice (Order Phthiraptera)

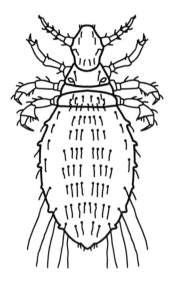

Figure 5.5
Little blue louse, Solenoptes capillatus. *(Illustration courtesy of Sarah Cox, Purdue University.)*

Veterinary Importance: The little blue louse is usually seen in clusters on the muzzle, dewlap, shoulders, back, and tail of mature animals. In heavily infested animals, aggregations of this louse often surround the eyes of their host. Like other cattle lice, the little blue louse can cause noticeable irritation and restlessness in cattle, resulting in skin abrasions and hair loss from scratching. It has been linked to decreases in milk production and hide damage.

Hosts: Cattle; and on a variety of ungulates

Description: The little blue louse is the smallest of the blood-sucking lice found on cattle (1.5 mm long). It has a short broad head, and the sides of the abdomen have tubercles projecting from the sides of each segment.

Biology/Behavior   Adult females deposit one to two eggs per day. Eggs hatch in about 10 to 12 days. The three nymphal instars develop to adults in about 11 days. The entire lifecycle takes approximately 28 days to complete.

LICE OF SWINE

Hog Louse (Suborder Anoplura; Family Haematopinidae) (Figure 5.6)

Species Name: *Haematopinus suis* (Linnaeus)

Geographic Distribution: Worldwide; most prevalent in temperate regions

Figure 5.6
Hog louse, Haematopinus suis.

Veterinary Importance: Hog lice will infest pigs of any age or condition; but unthrifty, malnourished animals are more susceptible to attack. Hog lice feeding activity can result in noticeable irritation and discomfort to swine. Significant blood loss can occur, especially in young animals. Infested animals scratch vigorously, rubbing against feeders, posts, or other objects. This can result in extensive hair loss and skin abrasions leaving the skin cracked, tender, and sore. Heavily infested animals become restless and become less profitable due to reduced performance. Skin abrasions can result in hide loss and discounted carcasses. Infested swine also become more susceptible to disease. Hog lice have also been shown to be capable of harboring and transmitting various disease agents. They can vector the virus that causes swine pox, a serious and potentially fatal disease in swine. Hog lice have also been implicated in the transmission of the causative agents of eperythrozoonosis, hog cholera, and the virus causing African swine fever.

Hosts: Swine

Description: The hog louse is the largest blood-sucking louse infesting domestic animals, reaching up to 6.4 mm long. It has a slate-blue body with brown-black markings.

Biology/Behavior: Hog lice are continuous obligatory parasites passing their entire lifecycle on their host. If dislodged from their host, hog lice rarely live for more than 3 days. They are more noticeable on swine in cooler weather. Hog lice feed mainly on tender areas of the skin around the host's neck, folds of skin behind the ears, and inside surfaces of legs and around the tail, often seen in clusters. They feed at frequent intervals throughout the day, feeding for 8 to 12 minutes at a time. Female hog lice lay from three to six eggs per day, attaching them firmly to a fair shaft (bristle) close to the skin. The whitish-colored eggs hatch in 2 to 3 weeks. The emerging nymphs feed for 2+ weeks, molting three times. Adult hog lice will live up to 5 weeks. There are typically 6 to 12 generations a year.

LICE OF SHEEP

Sheep Biting Louse (Mallaphaga; Suborder Ischnocera; Family Trichodectidae) (Figure 5.7)

Species Name: *Bovicola ovis* (Schrank)

Geographic Distribution: Worldwide

Veterinary Importance: The sheep biting louse is more prevalent on older sheep and animals in poor condition. They feed on skin scurf and can be very

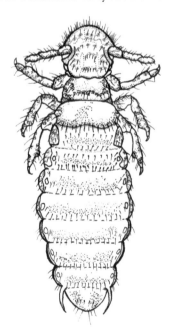

Figure 5.7
Sheep biting louse, Bovicola ovis. *(Copyright © Lincoln University, Natural Sciences Image Library.)*

irritating, causing sheep to rub vigorously against objects and to bite and pull at their wool. In heavily infested sheep, fleece becomes ragged and torn, often with large areas of wool being removed. Lice-infested sheep often have lower fleece yields and lower-quality wool.

Hosts: Sheep

Description: The sheep biting louse is a small louse 1.5 to 2 mm long. It is pale yellow in color with brown transverse stripes on the abdomen and a broad, red-brown head.

Biology/Behavior: Adult females will lay one to two eggs per day, cementing them individually to wool fibers near the skin. Eggs are laid on the legs in warmer temperatures and further up on the trunk of the body during colder weather. An individual female will produce up to 30 eggs in her lifetime. Eggs hatch after 9 to 25 days, depending on time of year and geographic location. Nymphs develop through three instars in about 3 weeks. Females mate within a few hours of molting to adults and start laying eggs within 3 to 4 days. The entire lifecycle takes 32 to 34 days. Sheep biting lice are more prevalent on the mid-dorsal line and body sides of sheep. Higher populations normally occur in cooler months. Transfer from one animal to another occurs when sheep are in close contact and more readily when sheep have short wool.

Sucking Lice on Sheep (Suborder Anoplura; Family Linognathidae)

Species Names:

African blue louse—*Linognathus africanus* (Kellogg and Paine)
Face and body louse—*L. ovillus* (Neumann) (Figure 5.8)
Sheep foot louse—*L. pedalis* (Osborn) (Figure 5.9)

Geographic Distribution: *Linognathus ovillus* and *L. pedalis* are found worldwide, and *L. africanus* is found in Africa, India, United States, and parts of Central America in semitropical areas.

Veterinary Importance: *Linognathus ovillus* prefers the face area where wool and hair meet, but in cold winter months they can be found on all body regions of sheep except the extremities of the lower limbs. *Linognathus pedalis* prefers areas of the feet, legs, and scrotum and tends to be found in clusters. Younger animals appear to be most susceptible to infestation. *Linognathus africanus* infestations can be found over all regions of the body. Both *L. ovillus* and *L. pedalis* rarely cause economic problems except for some irritation in heavier populations. With *L. africanus*, heavy infestations can cause considerable irritation and wool loss. Heavy infestations in deer have been reported to cause death.

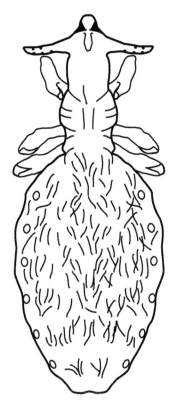

Figure 5.8
Face and body louse, Linognathus ovillus. *(Illustration courtesy of Sarah Cox, Purdue University.*

Hosts: *Linognathus ovillus* and *L. pedalis* on sheep; *L. africanus* on sheep, goats, and deer

Description: Typical of other *Linognathus* species, these three blood-feeding lice appear bluish in color and have narrow heads. They are about 2 mm long and abdomens have elongate bristles.

Biology/Behavior: The sheep foot louse has a lifecycle that spans an average of 43 days. Eggs incubate for an average of 17 days, and the three nymphal stages take a total of 21 days. After a 5-day preoviposition period, the adult females begin egg-laying activity. Unlike many other blood-sucking lice, the sheep foot louse can live up to 18 days off the host. This contributes to its successful transmission rate from one host animal to another.

The lifecycle of the face and body louse takes an average of 5 weeks to be completed. Adult females attach one egg per day to hair and wool on the face. Eggs incubate for 10 to 15 days, and nymphs develop through three instars in about 2 weeks. This louse can survive up to 4 days off the host.

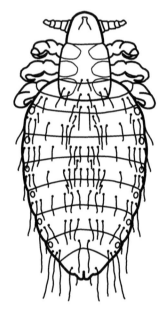

Figure 5.9
Sheep foot louse, Linognathus pedalis. *(Illustration courtesy of Sarah Cox, Purdue University.)*

The lifecycle of the African blue louse is similar to the face and body louse in duration. They may occur anywhere on the host's body, with emaciated and weak animals being most susceptible to lice buildup.

LICE OF GOATS

Chewing Lice (Mallaphaga; Suborder Ischnocera; Family Trichodectidae)

Species Names:

> Goat biting louse—*Bovicola caprae* (Gurlt) (Figure 5.10)
> Angora goat biting louse—*B. limbatus* (Gervais) (Figure 5.11)

Sucking Lice (Suborder Anoplura; Family Linognathidae)

Species Names:

> Goat sucking louse—*Linognathus stenopsis* (Burmeister) (Figure 5.12)
> African blue louse—*L. africanus* (Kellogg and Paine)

Geographic Distribution: *Bovicola caprae, B. limbatus,* and *L. stenopsis* are found worldwide, and *L. africanus* is found in Africa, India, United States, and parts of Central America in semitropical areas.

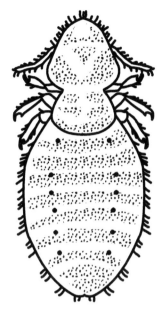

Figure 5.10
Goat biting louse, Bovicola caprae. *(Illustration courtesy of Sarah Cox, Purdue University.)*

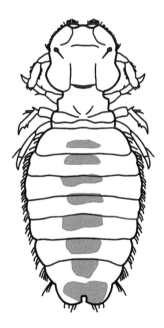

Figure 5.11
Angora goat biting louse, Bovicola limbatus. *(Illustration courtesy of Sarah Cox, Purdue University.)*

Figure 5.12
Goat sucking louse, Linognathus stenopsis. *(Illustration courtesy of Sarah Cox, Purdue University.)*

Veterinary Importance: The chewing lice, *B. caprae* and *B. limbatus*, feed mainly on skin scurf, hair, and detritus. They irritate goats, causing them to itch and rub against objects. This can be especially damaging to goat fiber, decreasing the amount of mohair and cashmere fiber produced and reducing fiber market value. Goat skin quality can also be affected.

Sucking lice can reduce weight gains and cause anemia in heavier infestations. Although heavy infestations generally develop only in goats in poor health or under stress, infestations can result in stunted weaned goats and occasional death in kids.

Hosts:

>*B. caprae*—goats, sheep
>*B. limbatus*—Angora goats
>*L. stenopsis*—goats
>*L. africanus*—sheep, goats, deer

Description: Chewing lice on goats, *B. caprae* and *B. limbatus*, are small lice ranging from 1 to 2 mm long. They have brownish heads and thorax.

Their abdomens are light brown to cream colored crossed with distinct red-brown bands. The two sucking lice species, *L. stenopsis* and *L. africanus*, are about 2 mm long, bluish gray in color, with narrow heads and abdomens with elongate bristles.

Biology/Behavior: For both species of chewing lice on goats, *B. caprae* and *B. limbatus*, their complete lifecycles take from 4 to 5 weeks to complete. Eggs hatch in about 7 to 12 days, and the three nymphal instars take from 14 to 30 days. The preoviposition for adult females averages 4 to 8 days. Life span for adults varies considerably ranging from 5 days to over 50 days.

   Sucking lice on goats complete their lifecycle in 4 to 5 weeks. Eggs hatch in 1 to 2 weeks, and the three nymphal stages develop over 2 to 3 weeks. Both *L. stenopsis* and *L. africanus* will be found widely dispersed over their host's body.

LICE OF DOMESTIC FOWL

Chicken Body Louse (Mallophaga; Suborder Amblycera; Family Menoponidae) (Figure 5.13)

Species Name: *Mencanthus stramineus* (Nitzsch)

Geographic Distribution: Worldwide

Figure 5.13 (see color insert following page 132)
Chicken body louse, Mencanthus stramineus.

Veterinary Importance: The chicken body louse is the most common and economically important louse found on poultry. Heavy infestations cause severe irritation to birds resulting in inflammation and scab development on the skin. These lice feed on barbs and barbules of the feathers and often puncture the quills and skin. When the skin and soft quills bleed from lice feeding activity, these lice will readily feed upon the oozing blood. This louse can cause feather loss, weight reduction, and reduced egg production and can cause death in younger, weaker birds. Populations can reach over 30,000 lice per host. They are most commonly seen on the breast, thighs, and vent area.

Hosts: Found on a variety of domestic and wild fowl, including chickens, turkeys, guinea fowl, pea fowl, quail, pheasants, ducks, and geese

Description: This chewing louse is relatively large with adults reaching 3.5 mm in length. The abdomen has a dense covering of setae.

Biology/Behavior: The entire lifecycle occurs on the host and requires from around 9 days to 2 to 3 weeks from egg to adult. Eggs are glued to the base of feathers in clusters and hatch in 4 to 7 days. Each of the three nymphal instars lasts around 3 days. Adult females will live up to 2 weeks with peak oviposition occurring at 5 to 6 days old.

Shaft Louse (Mallophaga; Suborder Amblycera; Family Menopnidae) (Figure 5.14)

Species Name: *Menopon gallinae* (Linnaeus)

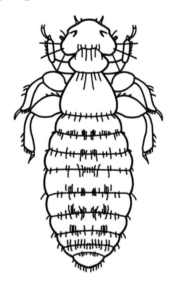

Figure 5.14
Shaft louse, Menopen gallinae. *(Illustration courtesy of Sarah Cox, Purdue University.)*

Geographic Distribution: Worldwide

Veterinary Importance: The shaft louse primarily feeds on the barbs of feathers and is found on the thigh and breast areas. It is rarely of serious concern but can cause death in younger birds in heavy infestations.

Host: Chickens, turkeys, and ducks

Description: The shaft louse is pale yellow in color and 1.7 to 2 mm in length. The abdomen has a sparse covering of small to medium long setae.

Biology/Behavior: Adult females lay eggs in clusters at the base of feathers located at the thigh and breast areas of their host. The complete lifecycle from egg to adult takes from 2 to 4 weeks. These lice are quite mobile and move rapidly on their host.

Chicken Head Louse (Mallaphaga; Suborder Ishnocera; Family Philopteridae) (Figure 5.15)

Species Name: *Cuclotogaster heterographus* (Nitzsch)

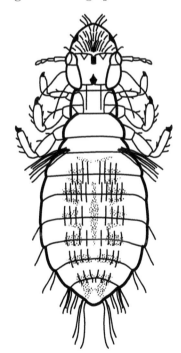

Figure 5.15
Chicken head louse, Cuclotogaster heterographus. *(Illustration courtesy of Sarah Cox, Purdue University.)*

Geographic Distribution: Worldwide

Veterinary Importance: The chicken head louse is usually found on skin and feathers of the head and neck of their host where they feed on skin, tissue, debris, and feathers. Younger birds are usually more affected than older birds. Occasionally, heavy infestations can lead to death.

Hosts: Found on a variety of domestic and wild fowl

Description: This louse is about 2.5 mm long with a rounded body and large round head. Three elongate bristles project from each side of the head.

Biology/Behavior: Adult females attach eggs singly to the base of downy feathers. Eggs hatch in 5 to 7 days, and nymphs pass through three instars, each lasting 6 to 14 days. The complete lifecycle from egg to adult takes around 35 days.

Wing Louse (Mallaphaga; Suborder Ischnocera; Family Philopteridae) (Figure 5.16)

Species Name: *Lipeurus caponis* (Linnaeus)

Geographic Distribution: Worldwide

Veterinary Importance: The wing louse is elongate and narrow with the adult being around 2.2 mm long and 0.3 mm wide. The hind legs are about twice as long as the front two pairs. There are small angular projections on the head in front of the antennae.

Biology/Behavior: Adult females attach eggs to the base of the feathers. Eggs hatch in 4 to 7 days. Nymphs develop through three instars in 20 to 40 days. The complete lifecycle from egg to adult takes around 35 days. Adults are slow moving and can live up to 36 days.

Other Lice Found on Domestic Fowl (Mallaphaga; Suborder Ischnocera; Family Philopteridae)

Species Names:

> Fluff louse—*Gonicocotes gallinae* (De Geer) (Figure 5.17)
> Large turkey louse—*Chelopistes meleagridis* (Linnaeus) (Figure 5.18)
> Large chicken louse—*Goniodes gigas* (Taschenberg) (Figure 5.19)

Veterinary Importance: These lice are sometimes found on domestic fowl but generally are not of economic importance.

Figure 5.16
Wing louse, Liperus caponis. *(Photo reproduced with permission by Marcelo de Campos Pereira, University of São Paulo, Brazil.)*

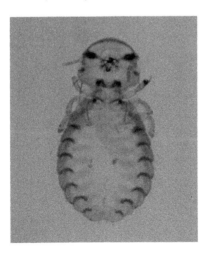

Figure 5.17
Fluff louse, Gonicocotes gallinae. *(Photo reproduced with permission by Marcelo de Campos Pereira, University of São Paulo, Brazil.)*

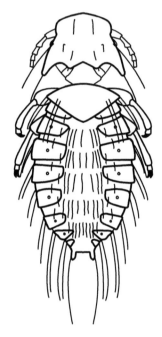

Figure 5.18
Large turkey louse, Chelopistes meleagridis. *(Illustration courtesy of Sarah Cox, Purdue University.)*

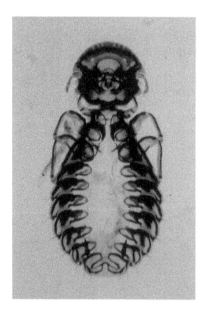

Figure 5.19
Large chicken louse, Goniodes gigas. *(Photo reproduced with permission by Marcelo de Campos Pereira, University of São Paulo, Brazil.)*

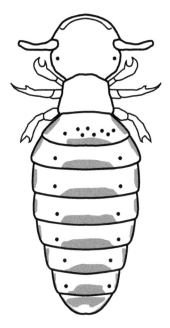

Figure 5.20
Horse biting louse, Bovicola equi. *(Illustration courtesy of Sarah Cox, Purdue University.)*

LICE OF EQUINE

Horse Biting Louse (Mallaphaga; Suborder Ischnocera; Family Trichodectidae) (Figure 5.20)

Species Name: *Bovicola equi* (Denny)

Geographic Distribution: Worldwide

Veterinary Importance: The horse biting louse is considered the most important louse of equine worldwide. They typically infest the side of the neck, shoulder, flanks, and base of the tail where they feed on the skin and hair of their host. In heavier infested animals, pruritus, hair loss, and coat damage may occur giving the animal a rough, unkempt appearance. Severely infested animals become nervous and irritable and typically rub against objects, kicking and stamping.

Hosts: Horses, mules, and donkeys

Description: The horse biting louse is about 2.5 mm long. It is chestnut brown in color with a lighter yellow abdomen with transverse dark stripes.

Biology/Behavior: Adult females oviposit eggs on finer hairs, avoiding the coarse hair of the mane and tail. There is a 7-day incubation of the eggs. The complete lifecycle from egg to adult takes from 27 to 30 days.

Horse Sucking Louse (Suborder Anoplura; Family Haematopinidae) (Figure 5.21)

Species Name: *Haematopinus asini* (Linnaeus)

Distribution: Worldwide, more prevalent in temperate areas

Veterinary Importance: The horse sucking louse is most prevalent on the head, neck, back, and inner surface of the thighs. Although not considered as important, it is often found more frequently than infestations of the horse biting louse. In heavy infestations, this blood-sucking louse can cause significant irritation to its host. Animals become restless and lose condition. Animals take on an unkempt appearance. Damage to the hair coat can result from animals rubbing and scratching against objects. Anemia may also develop from heavy infestations.

Hosts: Horses, mules, donkeys, and zebras

Description: The horse sucking louse is about 3 mm long. It is slate gray in color with a broad abdomen and narrow head.

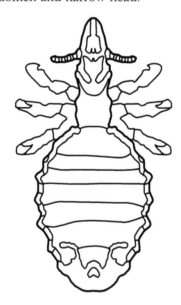

Figure 5.21
Horse sucking louse, Haematopinus asini. *(Illustration courtesy of Sarah Cox, Purdue University.)*

Biology/Behavior: Adult females oviposit from 50 to 100 eggs during their life span of 4 to 5 weeks. There is a 4- to 14-day incubation period. The three nymphal stages take 11 to 12 days to complete their development to adults. The complete lifecycle from egg to adult takes from 4 to 5 weeks. These lice have been observed to survive off their host for 2 to 3 days.

LICE OF CATS AND DOGS

Cat Biting Louse (Mallaphaga; Suborder Ischnocera; Family Trichodectidae) (Figure 5.22)

Species Name: *Felicola subrostrata* (Burmeister)

Geographic Distribution: Worldwide

Veterinary Importance: The cat biting louse is the only species of lice that commonly occurs on cats. It is not common on healthy cats but sometimes is found on elderly or chronically ill or poorly nourished cats. It can occur anywhere on the body. Heavy infestations can cause severe itching. Infested cats continually scratch resulting in damaged skin and hair coats. Infested cats also do not sleep well.

Hosts: Domestic and wild cats

Description: The cat biting louse is about 1.3 mm long. It has a distinctive triangular and anteriorly pointed head. Unlike many other Mallaphaga, the abdomen is smooth with few setae.

Biology/Behavior: This louse has a lifecycle similar to other lice with eggs attached to hair close to the skin and developing through three nymphal instars. The complete lifecycle from egg to adult takes around 30 to 40 days.

Dog Biting Louse (Mallaphaga; Suborder Ischnocera; Family Tricodectidae) (Figure 5.23)

Species name: *Trichodectes canis* (De Geer)

Geographic Distribution: Worldwide

Veterinary Importance: The dog biting louse, while not common on healthy animals, can sometimes build up heavy infestations on poorly nourished dogs. It can be harmful, particularly on puppies. It is most frequently found on the head, neck, and tail. It is quite active on dogs. Lice feeding activity on the skin can produce intense irritation, and skin damage can occur from aggressive

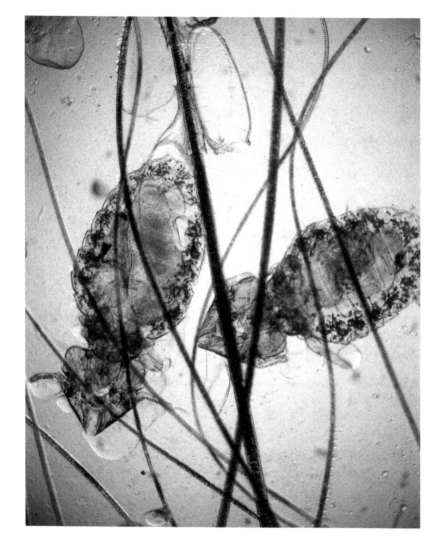

Figure 5.22
Cat biting louse, Felicola subrostrata. *(Photo reproduced with permission by Michael Dryden, Kansas State University.)*

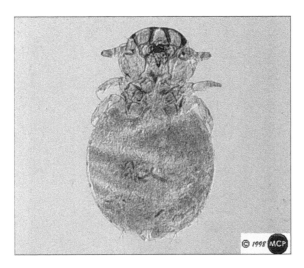

Figure 5.23
Dog biting louse, Trichodectes canis. *(Photo reproduced with permission by Marcelo de Campos Pereira, University of São Paulo, Brazil.)*

scratching. Inflammation and secondary bacterial infection can complicate the damage. This louse can also serve as the intermediate host of the double-poured tapeworm *Diplidium caninum*. Lice become infected when they ingest viable *D. caninum* eggs from dried host feces. Inside the louse, the tapeworm develops into a cysticercoid stage. If a dog ingests lice during grooming, the adult tapeworms will develop and emerge in the host's gut.

Hosts: Dogs and closely related canids including coyotes, foxes, and wolves

Description: The dog biting louse is about 1.5 mm long. The head is broader than long and the abdomen is covered with numerous large, thick setae.

Biology/Behavior: The lifecycle of this louse is similar to related species infesting animals. Eggs are attached to the hair, and there are three nymphal instars. The entire lifecycle from egg to adult takes from 3 to 5 weeks.

Dog Sucking Louse (Suborder Anoplura; Family Linognathidae) (Figure 5.24)

Species name: *Linognathus setosus* (Olfers)

Geographic Distribution: Worldwide, more common in cooler, temperate regions

Veterinary Importance: The dog sucking louse generally causes little irritation except in heavy infestations. Infestations are primarily found on the

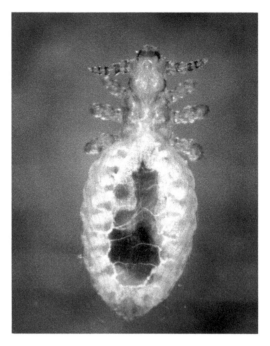

Figure 5.24
Dog sucking louse, Linognathus setosus. *(Photo reproduced with permission by Michael Dryden, Kansas State University.)*

neck and shoulders, frequently under a collar. As with the dog biting louse, heavy infestations can be irritating causing the host to rub, scratch, and bite at infested areas. This can result in sleeplessness, nervousness, alopecia, and a rough matted coat. Infestations are more prevalent in very young, old, or debilitated animals or animals kept in unsanitary environments.

Hosts: Dogs and closely related canids

Description: This louse is around 1.5 mm long with short antennae and head narrower than the thorax.

Biology/Behavior: Adult females will oviposit several eggs on the base of hair daily up to 30 during their life. The eggs hatch in 1 to 2 weeks. The nymphal stage develops through three instars over a 2- to 3-week period and then they molt into adults. The complete lifecycle from egg to adult takes from 3 to 6 weeks.

FLEAS (ORDER SIPHONAPTERA)

TABLE OF CONTENTS

The order Siphonaptera is a relatively small order of insects with approximately 2500 known species. Over 95% of these species are ectoparasites of mammals. There are a few species of fleas that infest avian hosts. Adult fleas are unique in appearance as compared to other insects. They are small (1 to 8 mm long), wingless, bilaterally compressed, and heavily chitinized. Many species are equipped with numerous bristles/spines called ctenidia in a comb-like appearance. These combs occur either on the ventral margin of the head (genal ctenidia) and/or the posterior margin of the prothorax (pronotal ctenidia). Most species have long legs adapted for jumping. At the ends of the legs are strong claws for clinging to and moving about a host. Fleas have well-developed mouthparts for piercing and sucking for taking blood meals. Both adult females and males feed on blood.

Fleas have both a parasitic stage in which adults are found on the host and a nonparasitic stage in which larvae feed off of their host (Figure 6.1). Adults mate on the host. Females can produce several hundred eggs during their lifetime, ovipositing a few eggs at a time. Eggs are usually laid on the host or sometimes in close proximity to the host. When on the host, they soon fall off onto the ground or host nest. Eggs hatch within a few days of oviposition.

Flea larvae are a light creamy color and are covered with hairlike bristles. They have small hooks present on the last body segment. They have a distinct head with well-developed chewing mouthparts. Larvae feed on a variety of proteinaceous organic debris including hair, skin, scales, scabs, dried host blood, feathers, and even cast exoskeletons and feces of adult fleas.

Most flea species pass through three larval instars over a 1- to 4-week period. Developmental time will vary, depending on available food, temperature, and moisture in their environment. They are quite sensitive to moisture and require a relative humidity of 60% to 80% for survival. They often die when relative humidity dips below 60%.

Upon maturity, a full-grown flea larva will spin a silken cocoon interwoven with debris from the habitat surrounding larval development. The duration of the pupal stage will vary from a few days to several months depending on environmental conditions and host availability. Adult emergence from the cocoon is usually triggered by vibrations caused by host movements. The longevity of adult fleas in the absence of available hosts can last for several months depending on environmental conditions. Survival is usually longer at lower temperatures and high humidity.

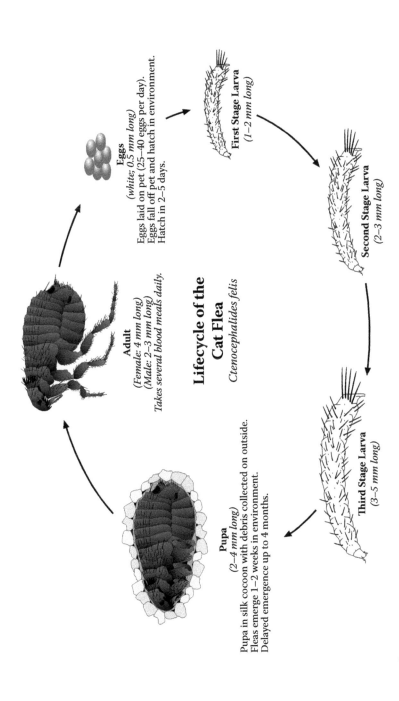

**Lifecycle of the
Cat Flea**
Ctenocephalides felis

Eggs
(white; 0.5 mm long)
Eggs laid on pet (25–40 eggs per day).
Eggs fall off pet and hatch in environment.
Hatch in 2–5 days.

First Stage Larva
(1–2 mm long)

Second Stage Larva
(2–3 mm long)

Third Stage Larva
(3–5 mm long)

Adult
(Female: 4 mm long)
(Male: 2–3 mm long)
Takes several blood meals daily.

Pupa
(2–4 mm long)
Pupa in silk cocoon with debris collected on outside.
Fleas emerge 1–2 weeks in environment.
Delayed emergence up to 4 months.

Figure 6.1
Flea lifecycle. (Illustration courtesy of Scott Charlesworth, Purdue University, based in part on R. E. Elbel, 1991. In: Immature Insects, Vol. 2.)

Both adult male and female fleas feed on blood from their host. Individual blood meals are small, but repeated feeding and high infestations can cause significant blood loss. Flea bites are irritating and can cause inflammation and pruritus at the flea bite site. Additional damage often occurs from scratching and self-biting by the host animal. Flea bites can cause allergic reactions in their host. Also, depending on the species, fleas are known vectors or intermediate hosts of several pathogens, including viruses, bacteria, rickettsiae, and helminthes.

CAT FLEA (ORDER SIPHONAPTERA; FAMILY PULICIDAE) (FIGURE 6.2)

Species Name: Ctenocephalides felis *(Bouche)*

Geographic Distribution: Worldwide

Veterinary Importance: The cat flea is the most common flea found on domestic cats and dogs. It also feeds on humans and several other mammals. It can cause extensive irritation resulting in skin inflammation and dermatitis. Young animals can become anemic with heavy infestations. Flea allergy dermatitis

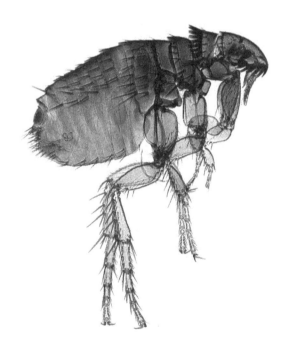

Figure 6.2
Cat flea, Ctenocephalides felis. *(Photo reproduced with permission by Steve Upton, Kansas State University.)*

can cause strong grooming reactions by the host animal. Allergically infested animals continuously lick, scratch, and bite at flea bite areas, causing hair loss and skin lesions. Secondary infection is also common at bite site lesions.

The cat flea serves as an intermediate host for the double-pored tapeworm, *Dipylidium caninum*. Tapeworm eggs are passed out in the host feces and ingested by flea larvae. These eggs hatch inside the flea larvae. The tapeworm larvae develop in the flea through the flea's larval, pupal, and adult development. In the adult flea, the tapeworm encapsulates as a cysticercoid. When the infected flea is ingested by a host grooming itself, adult tapeworms emerge in the host digestive tract.

Host: Dogs, cats, and several other mammals, including humans

Description: The cat flea has both genal and pronotal ctenidia. The heads of adult females and males are elongate anteriorly as compared to more rounded in the dog flea, *C. canis*.

Biology/Behavior: Adult female cat fleas, once they find a suitable host, tend to stay on the host. Adult females will begin laying eggs within 1 to 4 days of their first blood meal. They will lay from 10 to 30 eggs per day and produce several hundred over their lifetime of 1 to 2 months. Eggs are usually deposited on the host but soon fall off onto the ground. Under optimal conditions, the eggs will hatch within 1 to 2 days. Their lifecycle follows the description as discussed earlier in this chapter.

DOG FLEA (ORDER SIPHONAPTERA; FAMILY PULICIDAE) (FIGURE 6.3)

Species Name: Ctenocephalides canis (Curtis)

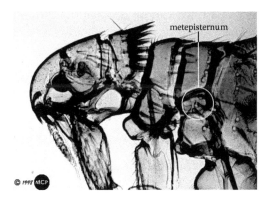

Figure 6.3
Dog flea, Ctenocephalides canis. *(Photo reproduced with permission by Marcelo de Campos Pereira, University of São Paulo, Brazil.)*

Geographic Distribution: Worldwide

Veterinary Importance: Although widely distributed throughout the United States and around the world, the dog flea is much less common than the cat flea. The cat flea is actually the most common flea found on dogs. The dog flea causes similar damage to its host, including host hypersensitivity and flea allergy dermatitis. And, like the cat flea, it serves as an intermediate host of the double-pored dog tapeworm.

Hosts: Dogs, cats, and several other mammals

Description: Like the cat flea, the dog flea has both genal and pronotal ctenidia. The heads of adult females and males are rounded anteriorly as compared to more elongate in the cat flea, *C. felis.*

Biology/Behavior: The lifecycle of the dog flea is essentially the same as that of the cat flea.

HUMAN FLEA (ORDER SIPHONAPTERA; FAMILY PULICIDAE) (FIGURE 6.4)

Species Name: **Pulex irritans** *(Linnaeus)*

Geographic Distribution: Cosmopolitan worldwide

Veterinary Importance: The human flea is of less veterinary importance than cat and dog fleas. However, it has been found as a pest species on swine. Heavily infested swine are often highly irritated and will scratch continuously when infested with the human flea.

Description: The human flea can be distinguished from cat and dog fleas in that it does not have genal or pronotal ctenidia. Also, it has a smoothly rounded head.

Biology/Behavior: The lifecycle of the human flea is similar to that of the cat flea.

ORIENTAL RAT FLEA (ORDER SIPHONAPTERA; FAMILY PULICIDAE) (FIGURE 6.5)

Species Name: **Xenopsylla cheopis** *(Rothchild)*

Geographic Distribution: Cosmopolitan worldwide, wherever roof rats, *Rattus rattus,* occur

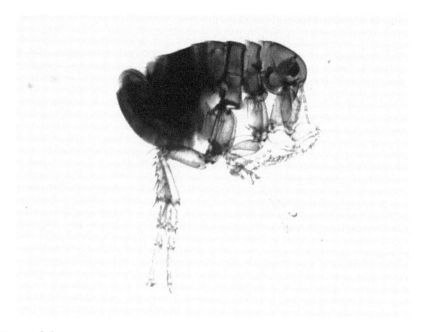

Figure 6.4
Human flea, Pulex irritans.

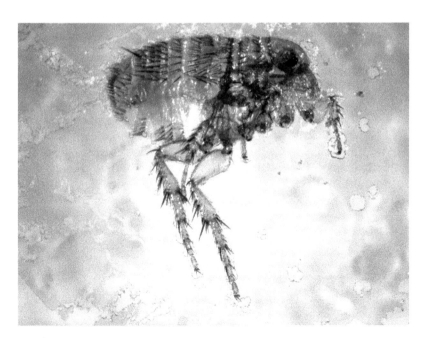

Figure 6.5
Oriental rat flea, Xenopsylla cheopis.

Veterinary Importance: The primary host of the oriental rat flea is the roof rat, *Rattus rattus*, although it does occur frequently on the Norway rat and house mouse. It is not of significant veterinary importance. However, if its rodent hosts become scarce, they will feed on humans, dogs, cats, chickens, and other hosts. Of most concern is that it is a chief vector of the pathogen *Yersinia pestis* that causes plague in humans. It is also an inter-mediate host of helminthes, including *Hymenolepis diminuta*, *H. nana*, and *Trichinella siralis*.

Hosts: Rats and other rodents primarily. It will feed on humans, cats, dogs, chickens, and other animals.

Description: The oriental rat flea is similar in appearance to the human flea with both genal and pronotal ctenidia absent and head smoothly rounded.

Biology/Behavior: The lifecycle of the oriental rat flea is similar to that of the cat flea.

STICKTIGHT FLEA (ORDER SIPHONAPTERA; FAMILY PULICIDAE) (FIGURE 6.6)

Species Name: Echidnophaga gallinacea *(Westwood)*

Geographic Distribution: Worldwide in warmer climates

Veterinary Importance: The sticktight flea is primarily of importance as a parasite of birds. It can be a serious pest of chickens. In the United States, infestations are commonly reported in the southern regions of the country, primarily in subtropical areas. On poultry, infestations can occur around the head where the adult flea firmly attaches its mouthparts into the host skin. Clusters of these fleas can often be found around the eyes, comb, wattles, and other bare spots. They can cause skin irritation and ulceration. Bird egg production and feed efficiency will decline in heavier infestations. Severe infestations can lead to blindness and even death in young birds.

Hosts: Domestic poultry. It can also be found on a variety of wild birds and on mammals including dogs, cats, horses, deer, and a variety of smaller mammals.

Description: The sticktight flea is small (2 mm long) with the head sharply angled and squarish in front. There are no genal or pronotal ctenidia.

Biology/Behavior: Newly emerged females actively seek a host and then, once found, will aggregate on the head or other bare spots to firmly attach their mouthparts into the skin. Females mate in this position. They will remain

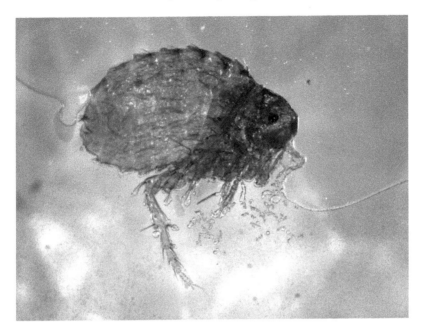

Figure 6.6
Sticktight flea, Echidnophaga gallinacea.

attached for 2 to 6 weeks. Eggs are either laid in the ulcerated attachment site or drop to the ground. If larvae hatch on the host, they soon drop to the ground to complete their development. Larvae feed on host feces and other organic debris. The entire lifecycle takes 1 to 2 months.

CHIGOE (ORDER SIPHONAPTERA; FAMILY TUNGIDAE) (FIGURE 6.7)

Species Name: Tunga penetrans (Linnaeus)

Geographic Distribution: Tropical and subtropical regions of North and South America, West Indies, and Africa

Veterinary Importance: The chigoe is known by a variety of names including jigger, chique, and sand flea. It is more frequently encountered on humans where it usually attacks the feet. Resultant sores may fill with pus and become infected. It may be found on swine where it infests the feet, snout, scrotum, or teats. When found on teats, it can prevent adequate milk flow for proper nourishment of suckling pigs.

Hosts: Humans and some domestic animals, such as swine

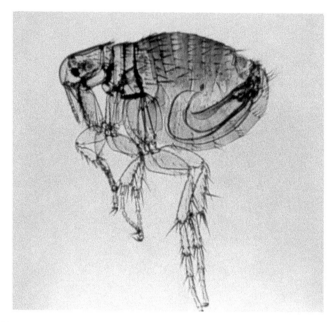

Figure 6.7
Chigoe, Tunga penetrans.

Description: This flea is quite small (1 mm long), but gravid females may swell to the size of a pea. The head is angular, and it lacks genal and pronotal ctenidia. The thoracic segments are narrow at the top.

Biology/Behavior: The adult female burrows into the skin, commonly at the feet of its host, where it remains attached. A nodular swelling surrounds the female. She feeds on host fluids, expanding to about the size of a pea in 8 to 10 days. There remains an opening in the nodule through which the female respires, mates with a free-living male, and expels her eggs. Eggs drop to the ground and develop through two larval instars. The total lifecycle from egg to adult takes from 3 to 6 weeks.

NORTHERN RAT FLEA (ORDER SIPHONAPTERA; FAMILY CERATOPHYLLIDAE)

Species Name: Nosopsyllus fasciatus (Bosc)

Geographic Distribution: Widespread in Europe and North America and other temperate regions of the world

Veterinary Importance: Primarily found on the Norway rat, *Rattus norvegicus*, it occasionally parasitizes other rodents and domestic mammals and humans. It has been shown to transmit several zoonotic pathogens such as the disease agents of plague and murine typhus. It is generally considered not to be of veterinary importance.

Hosts: Primarily found on the Norway rat and other related rodents

Description: The northern rat flea has a pronotal ctenidium and lacks a genal ctenidium. The body is elongated and is about 3 to 4 mm in length.

Biology/Behavior: The lifecycle of the northern rat flea is similar to that of the cat flea.

EUROPEAN CHICKEN FLEA (ORDER SIPHONAPTERA; FAMILY CERATOPHYLLIDAE) (FIGURE 6.8)

Species Name: **Ceratophyllus gallinae** *(Schrank)*

Geographic Distribution: Worldwide where domestic poultry are raised

Veterinary Importance: The feeding activity of this flea on poultry can cause irritation, restlessness, and anemia. They sometimes can become a nuisance to humans working in infested poultry facilities or in dwellings invaded by infested wild birds. It is often found in wild bird nesting material.

Hosts: Poultry and numerous wild birds. It has been found associated with more than 75 species of wild birds.

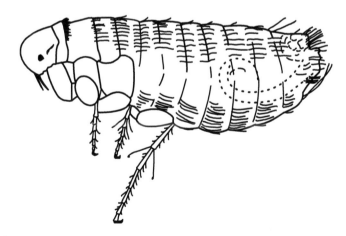

Figure 6.8
European chicken flea, Ceratophyllus gallinae. *(Illustration courtesy of Sarah Cox, Purdue University.)*

Description: This flea has a pronotal ctenidium and lacks a genal ctenidium. The body is elongate and about 4 mm long.

Biology/Behavior: The lifecycle is similar to that of the cat flea but has adapted to correspond to the migration of many passerine birds. It overwinters in its cocoon and emerges when spring temperatures rise and host birds settle on the nest.

WESTERN CHICKEN FLEA (ORDER SIPHONAPTERA; FAMILY CERATOPHYLLIDAE) (FIGURE 6.9)

Species Name: **Ceratophyllus niger** *(Fox)*

Geographic Distribution: Western North America

Veterinary Importance: This flea has been reported to be common in poultry houses in the western United States and Canada, where it sometimes causes an annoyance to domestic birds.

Hosts: Primarily on poultry, but occasionally feeds on rats, cats, dogs, and humans

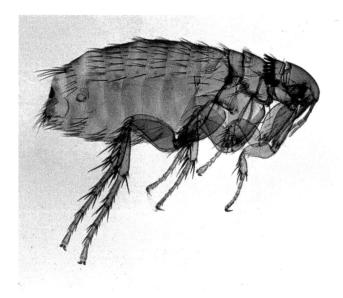

Figure 6.9
Western chicken flea, Ceratophyllus niger. *(Photo reproduced with permission by Steve Upton, Kansas State University.)*

Fleas (Order Siphonaptera)

Description: It is similar in appearance to the European chicken flea.

Biology/Behavior: The lifecycle is similar to that of the cat flea.

Figure 2.5
Pteromalid wasp. (Photo reproduced with permission by Jim Kalisch, University of Nebraska.)

Figure 3.10
Horse fly, Tabanus atratus. (Photo reproduced with permission by Jim Kalisch, University of Nebraska, Lincoln.)

Figure 3.24
Face flies on cow.

Figure 3.25
House fly, Musca domestica. (Photo reproduced with permission by Jim Kalisch, University of Nebraska, Lincoln.)

Figure 4.3
Common cattle grub adult, Hypoderma lineatum.

Figure 5.13
Chicken body louse, Mencanthus stramineus.

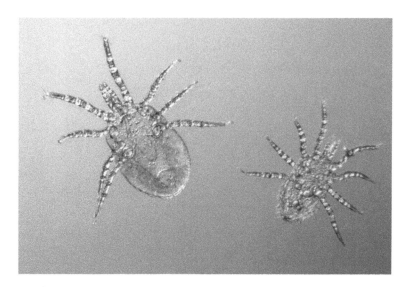

Figure 7.4
Chicken mite, Dermanysus gallinae. *(Photo reproduced with permission by Jim Kalisch, University of Nebraska.)*

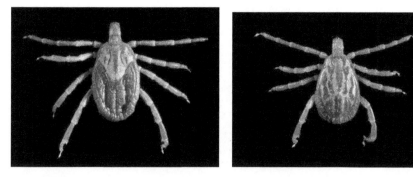

Figure 8.4
(Top) Gulf Coast tick female, Amblyomma maculatum. *(Bottom) Gulf Coast tick male,* Amblyomma maculatum. *(Photos reproduced with permission by James Gathany, Centers for Disease Control and Prevention.)*

CHAPTER 7

MITES

TABLE OF CONTENTS

Of the some 30,000 identified species of mites, most known species are free living, predaceous, or plant feeders. However, there are around 250 species of mites recognized as being of public health and/or veterinary importance. Mites belong to the subclass Acari in the class Arachnida. Within this subclass, in addition to the order Ixodida, which includes the ticks covered in Chapter 8, the orders of mites of veterinary importance include Mesostigmata, Prostigmata, and Astigmata (Table 7.1).

GENERAL PATHOLOGY

Mites of veterinary importance will infest animals in one of three ways. Some mite species will feed externally on their host. Some feed subcutaneously, often with scabs forming over the infested areas. Other species invade their host internally within the lungs, air sacs, or various body canals. Mite infestation can result in various pathological effects to their host, including the following:

- Dermatitis—either temporary irritation of the skin due to mite feeding on host skin, fur, or feathers, or persistent dermatitis caused by tissue damage from mites invading the skin or hair follicles
- Mite-induced allergic reactions and cutaneous hypersensitivity
- Loss of blood or other body fluids
- Transmission of pathogens or intermediate hosting of parasites

Some mite species are not considered as pest species but actually are considered beneficial in that they are predatory on other arthropods. One species, *Macrocheles muscaedomesticae*, known as the house fly mite, is an effective predator of house flies and is discussed at the end of this chapter.

Table 7.1
Mites of Veterinary Importance

Species	Common Name
Order Mesostigmata	
Family Macronyssidae	
Ornithonyssus sylviarum (Canestrini and Fanzago)	Northern fowl mite
O. bursa (Berlese)	Tropical fowl mite
O. bacoti (Hirst)	Tropical rat mite
Family Dermanyssidae	
Dermanyssus gallinae (De Geer)	Chicken mite
Family Halarachnidae	
Rallietia auris (Leidy)	Cattle ear mite
Family Macrochelidae	
Macrocheles muscaedomesticae (Scapoli)	House fly mite
Family Uropodidae	
Fuscuropoda vegetans (De Geer)	
Order Prostigmata	
Family Trombiculidae	
Neoschongastia americana (Hirst)	Chigger
Trombicula (Eutrombicula) alfreddugesi (Oudemans)	Turkey chigger
T. (Eurtrombicula) splendens (Ewing)	Common chigger
Family Cheyletidae	
Cheyletiella blakei (Smiley)	Cat fur mite
C. parasitivorax (Megnin)	Rabbit fur mite
C. yasguri (Smiley)	Dog fur mite
Family Psorergatidae	
Psorobia ovis (Wormersley)	Sheep itch mite
P. bos (Johnston)	Cattle itch mite
Family Demodicidae	
Demodex bovis (Stiles)	Cattle follicle mite
D. canis (Leydig)	Dog follicle mite
D. caprae (Railliet)	Goat follicle mite
D. cati (Megnin)	Cat follicle mite
D. equi (Railliet)	Horse follicle mite
D. folliculorum (Simon)	Follicle mite
D. phylloides (Csoker)	Hog follicle mite

Continued

Table 7.1 (Continued)
Mites of Veterinary Importance

Species	Common Name
Order Astigmata	
Family Analgidae	
Megninia sp.	Feather mites
Family Mycoptidae	
Myocoptes musculinus (Koch)	Mycoptic mange mite
Family Psoroptidae	
Psoroptes ovis (Hering)	Sheep scab mite
P. cuniculi (Delafond)	Rabbit ear mite
Chorioptes bovis (Hering)	Chorioptic mange mite
Otodectes synotis (Hering)	Ear mite
Family Scaroptidae	
Sarcoptes scabiei (De Geer)	Sarcoptic mange mite, scabies mite, itch mite, hog mange mite, others
Notoedres cati (Hering)	Notoedric cat mite
Family Knemidokiptidae	
Knemidocoptes mutans (Robin and Lanquetin)	Scaly-leg mite
K. gallinae (Railliet)	Depluming mite
Family Laminosioptidae	
Laminosioptes cysticola (Vizioli)	Fowl cyst mite
Family Cytoditidae	
Cytodites nudus (Vizioli)	Airsac mite

GENERAL MORPHOLOGY (FIGURE 7.1)

Mites are quite small, usually less than 1 mm long. The body is saclike and shows no segmentation. It has two body regions: the anterior gnathosoma composing the mouthparts, and the posterior idiosoma comprising the rest of the body. The idiosoma is divided into the podosoma that includes the legs and the opisthosoma. Adult and nymphal mites possess four pairs of legs, and larvae have three pairs of legs. The mouthparts vary considerably between mites, but most parasitic species have chelicerae used to penetrate the skin of their host. Mites lack recurred hypostomal teeth found in ticks and used to permit firm attachment to their host. Most species have a tracheal respiratory system with the presence of spiracular openings. Tracheal systems and spiracular openings are lacking in the Astigmata which rely on cuticular oxygen exchange. Eyes are usually absent in mites. Many mites simply rely on hairs or setae with sensory function. Because the digestive system of mites primarily handles liquefied food, there are developed salivary

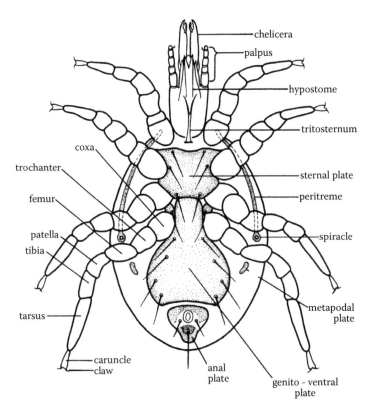

Figure 7.1
Generalized mite diagram. (Illustration courtesy of the Communicable Disease Center, U.S. Department of Health, Education, and Welfare, Harry D. Pratt.)

glands in the anterior portion of the idiosoma. Enzymes secreted in the saliva assist in preorally digesting food. In addition to digestive enzymes, these glands secrete anticoagulants in hematophagus mites, and in some mites (e.g., chiggers), they secrete cementing substances to assist in adhering mouthparts to host skin.

GENERAL LIFE HISTORY

Most mites have similar lifecycles. Eggs hatch into six-legged larvae. Larvae molt into eight-legged nymphs. There are usually from one to three nymphal stages termed protonymph, deutonymph, and tritonymph, respectively. In some mites, the deutonymph is an inactive, nonfeeding stage. Mature nymphs then molt into eight-legged adults. Developmental time from egg to adult varies considerably between mite species and is discussed under each species included in this chapter.

EXTERNAL FEEDING MITES

Northern Fowl Mite (Order Mesostigmata; Family Macronyssidae) (Figure 7.2)

Species Name: *Ornithonyssus sylviarum* (Canestrini and Fanzago)

Geographic Distribution: Worldwide in temperate regions

Veterinary Importance: The northern fowl mite is considered one of the most important, if not the most important, external parasites of poultry in North America, throughout Europe, and in New Zealand and Australia, and other parts of the world where it has been introduced. It is especially of economic concern in caged poultry operations. They are continuous obligatory parasites spending their entire life on their host. Mite infestations are most frequently observed around the vent area of birds. Light infestations often go unnoticed initially, but as mite numbers increase, vent area feathers become matted and take on a gray to black discoloration from larger mite numbers, mite feces and

Figure 7.2
Northern fowl mite, Ornithonyssus sylviarum.

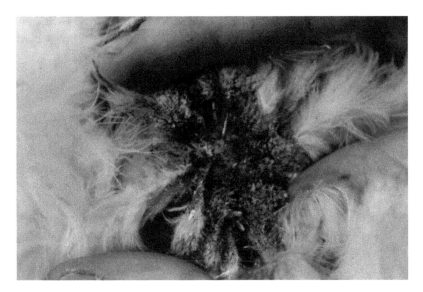

Figure 7.3
Northern fowl mite infestation.

cast exoskeltons, and mite eggs (Figure 7.3). In heavily infested birds, northern fowl mite numbers can reach over 25,000 per bird. The skin around the vent area where mites feed becomes scaly, scabby, and thickened. Heavy infestations can cause decreased egg production, weight loss, anemia, loss of condition, and sometimes death. In heavily infested laying hens, reductions as high as 15% have been reported in egg production. The greatest impact on body weight, feed efficiency, and egg production loss generally occurs when hens are infested with mites before they reach full egg production.

Northern fowl mites tend to be more abundant during cooler months and usually more abundant on younger birds between 20 and 30 weeks of age. Also, in caged layer facilities, there is a tendency for mites to be more abundant on birds caged alone than those caged in groups. Chickens are social by nature and are often more stressed when caged alone.

Northern fowl mites can be irritating to people and sometimes can cause dermatitis and allergic reaction in poultry farm workers. In addition to infesting chickens, these mites are also commonly found in nests of pigeons and various wild birds and have been reported as household pests when bird nests are proximate to human dwellings.

Although pathogens of poultry, such as viruses that cause Newcastle disease and fowlpox, have been recovered from northern fowl mites after feeding on infected birds, there is no evidence that these mites transmit these agents in nature.

Hosts: Poultry and many wild birds

Description: Adult female northern fowl mites are about 0.75 to 1 mm in length with relatively long legs. They vary in color from a light red to black. Individual mites can be easily seen with the naked eye.

Biology/Behavior: The entire lifecycle takes place on the host. Eggs are laid in masses at the base of the feathers that are primarily located around the vent area of the bird. Eggs hatch in about a day, and larvae molt to the proto-nymphal stage in less than a day without feeding. After the protonymphs feed on blood, they molt into nonfeeding deutonymphs. These then rapidly molt into adults. The lifecycle can be completed in 5 to 12 days, and under favorable conditions large populations can develop rapidly on birds.

Transmission of northern fowl mites from bird to bird and from cage to cage is usually from direct bird contact. Transmission between poultry facilities can occur in many ways. Poultry house workers going from one facility to another, egg crates and flats being moved from one facility to another, vehicles carrying chickens, wild birds entering poultry facilities, rat and mice movement, and infested pullets being placed in egg-layer facilities have all been implicated in the dissemination of this mite. Also, northern fowl mites have been shown to survive off the host for several days which enhances their being able to survive while being transported. This survival off the host also factors in when infested birds are removed from a poultry house and new birds are brought in. Subsequently, the new birds become infested, even after the house may sit empty for a few days.

Chicken Mite (Order Mesostigmata; Family Dermanyssidae) (Figure 7.4)

Species Name: *Dermanyssus gallinae* (De Geer)

Geographic Distribution: Cosmopolitan worldwide

Veterinary Importance: The chicken mite, also known as the chicken red mite, is an obligate parasite of wild and domestic birds. It occasionally infests mammals, including dogs, cats, horses, cattle, and rodents, that become associated with poultry facilities or infests bird nests. It will also bite people.

This mite can be a major problem in poultry operations, primarily on birds maintained on floor operations rather on caged laying hens, because it does not complete its lifecycle solely on the host. It is only present and feeds on birds at night. It has also been found to be a problem in brooder houses where nest boxes are used.

Chicken mite infestations can be debilitating, resulting in significant skin irritation and lesions (commonly seen on the breast and legs of birds), loss of vigor, stunted growth, feather loss, reduced egg production, anemia, and

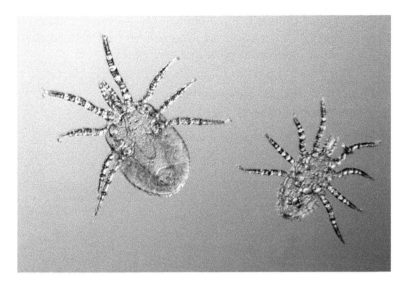

Figure 7.4 (see color insert following page 132)
Chicken mite, Dermanysus gallinae. *(Photo reproduced with permission by Jim Kalisch, University of Nebraska.)*

sometimes death, especially in younger birds. The chicken mite has been incriminated as a vector of various encephalitis viruses and disease agents responsible for fowl cholera and fowl spirochetosis. However, it is probably of only minor importance in such disease transmission under natural conditions.

In addition to infesting chickens, these mites are commonly found in and around nests of various wild birds and have been reported as household pests when bird nests are proximate to human dwellings.

Hosts: Poultry and many other birds. It can also feed on a variety of mammals, including humans, dogs, cats, horses, cattle, and rodents.

Description: Adult mites are 0.75 to 1 mm long with long legs. They are usually a grayish color, becoming red when engorged with blood. The chelicerae are distinctively elongate and whiplike.

Biology/Behavior: Adults mate off the host. After feeding to repletion on a host, at night the females will migrate to crack and crevice areas off the host to lay their eggs. Eggs are typically deposited in groups of four to seven eggs at a time at approximately 3-day intervals. From 20 to 24 eggs are produced by a female during her lifetime. Larvae hatch from eggs in 1 to 3 days and do not feed. Both protonymphs and deutonymphs require blood meals with both stages lasting approximately 1 to 2 days, depending on environmental conditions. Under suitable conditions, the entire lifecycle from egg to adult can be completed in 1 week. Adult mites can survive several months off the host without feeding. This enables them to survive for extended periods in abandoned bird nests and unoccupied poultry houses.

Tropical Fowl Mite (Order Mesostigmata; Family Macronyssidae) (Figure 7.5)

Species Name: *Ornithonyssus bursa* (Berlese)

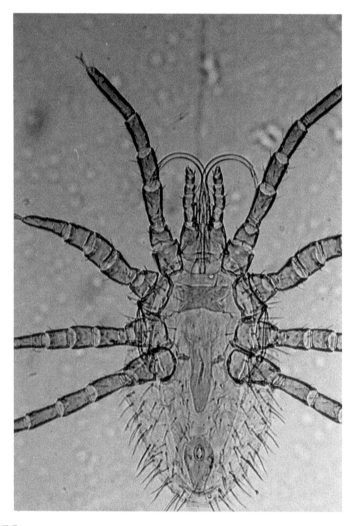

Figure 7.5
Tropical fowl mite, Ornithonyssus bursa. *(Photo reproduced with permission by Marcelo de Campos Pereira, University of São Paulo, Brazil.)*

Geographic Distribution: Worldwide throughout warmer regions

Veterinary Importance: This mite is a common external parasite of wild and domestic birds in tropical and subtropical climates around the world. However, it is seldom reported on poultry in the United States. Heavy infestations in chickens and other domestic poultry can result in anemia, decreased weight gain, reduced egg production, and sometimes death, more so in younger birds. Skin irritation caused by mite feeding will sometimes cause setting hens to leave their nests. Similar to northern fowl mite infestations, this mite is primarily found in feathers around the vent area of birds. Feathers become matted and dirty in appearance from the buildup of mites, cast exoskeletons, mite eggs, and mite feces.

Hosts: Poultry and many other birds

Description: The tropical fowl mite is similar in appearance to the northern fowl mite, *O. sylviarum.*

Biology/Behavior: The tropical fowl mite lays its eggs in the fluff of host feathers or occasionally in nesting materials off the host. Eggs hatch in about 2 to 3 days. Nonfeeding larvae molt to blood-feeding protonymphs in less than a day. The complete lifecycle from egg to adult takes about 1 week. It has been shown to survive off of a host for up to 10 days.

Chiggers (Order Prostigmata; Family Trombiculidae) (Figure 7.6)

Species Names:

Neoschongastia americana (Hirst)—Turkey chigger
Tombicula (Eutrombicula) alfreddugesi (Oudemans)—Common chigger, harvest mite
Trombicula (Eutrombicula) splendens (Ewing)
Others

Geographic Distribution

N. americana—United States into tropical areas of the Americas
T. alfreddugesi—Western Hemisphere from Canada to South America
T. splendens—Eastern United States from the Gulf Coast north to Massachusetts and Ontario, Canada, and west to Minnesota. It is more common in the southeastern United States.

Other species are found scattered throughout different geographic ranges around the world.

Veterinary Importance: There are more than 1500 species of mites in the family Trombiculidae, of which around 50 are known to be of public

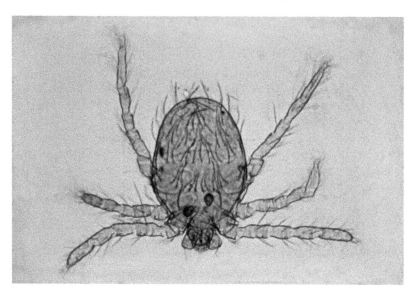

Figure 7.6
Chigger.

health and/or veterinary importance. The six-legged larval stage of these are parasitic. Nymphs and adults are free-living predators and primarily feed on small arthropods and their eggs. Most species of domestic animals are susceptible to chigger infestation. Cats, dogs, horses, and livestock kept outdoors are prone to being infested in geographic areas where chiggers are active.

When they find a host, chiggers do not burrow into the skin but attach to the epidermis and feed externally. They feed primarily on partially digested skin cells and lymph dissolved by their salivary secretions. They do not feed directly on blood. When they attach to a host, they form a feeding tube produced by the interaction of their saliva and surrounding host tissue. They usually only stay attached from 1 to 2 days, but the host reaction and irritation may persist for several days.

In most situations, chigger infestation produces mild pruritis. However, heavy infestations can result in severe itching and formation of scabby skin lesions. In mammals, chigger infestations can occur on the face, ears, neck, appendages, feet, and underside. Resulting skin lesions can be quite irritating, and skin damage can occur when the host animal rubs, scratches, and bites at affected areas. Domestic birds, such as chickens and turkeys, that are kept outdoors, often show reactions to chigger infestations. Chigger infestations can be found under the wings, around the vent, and on the head around the eyes and on combs and wattles. Noticeable irritation and dermatitis can result in more heavily infested birds. Unsightly lesions, when trimmed away during processing, reduce the cosmetic appeal and value of bird carcasses.

Hosts: Wide variety of animals including mammals, birds, reptiles, and amphibians

Description: Chiggers, being larval mites, have six legs. They are quite small (only 0.2 mm long). They are red to orange in color.

Biology/Behavior: Eggs are laid in the soil or ground debris. After around a 6-day incubation period, the eggs hatch into a nonfeeding prelarval stage. Feeding larvae (chiggers) become active within the next 6 days. When the larvae find a suitable host, they feed from 1 to 5 days, drop to the ground, and molt into the nymphal stage. After three nymphal instars, the adults emerge. Deutonymphs and adults are predatory, feeding on small arthropods and their eggs. The complete lifecycle takes from 2 to 12 months or longer, depending on the species and environmental conditions. Species in temperate regions generally have 1 to 3 generations per year. In warmer climates, year-round continuous development can occur.

Other External Feeding Mites

There are a few other species of external feeding mites of veterinary importance, including the following:

Ornithonyssus bacoti (Hirst)—tropical rat mite
Cheyletiella blakei (Smiley)—cat fur mite
C. parasitivorax (Megnin)—rabbit fur mite
C. yasguri (Smiley)—dog fur mite
Megninia sp.—feather mites
Myocoptes musculinus (Koch)—mycoptic mange mite

These mites may occasionally bite people and be found associated with domestic animals but are generally not of economic importance. Fur mites on cats and dogs sometimes warrant treatment, however.

SUBCUTANEOUS FEEDING OR SCAB-FORMING MITES

Sheep Scab Mite (Order Astigmata; Family Psoroptidae) (Figure 7.7)

Species Name: *Psoroptes ovis* (Hering)

Geographic Distribution: Cosmopolitan worldwide

Veterinary Importance: This mite can cause severe economic damage to sheep and cattle. It causes what has been termed *sheep scab* and *cattle scab.* Psoroptic mites actually feed on the surface of the skin and do not pierce the epidermis. They feed on lipids, skin cells, and skin secretions with their

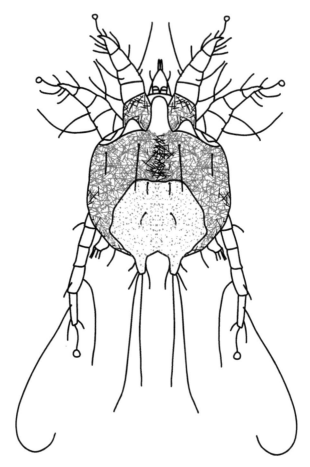

Figure 7.7
Sheep scab mite, Psoroptes ovis. *(Illustration courtesy of Sarah Cox, Purdue University.)*

sucking mouthparts. Host response to this feeding activity results in serous exudates forming a dry, yellow crust, surrounded by a border of inflamed skin covered in moist crust. Mites can be found on the moist tissue at the edge of these lesions. In untreated animals, these lesions will quickly expand and can develop over a major portion of the body. This can result in extensive wool and hair loss, reduced weight gains, and reduced feed efficiency. In some cases, death of the animal may result. The severity of response to mite infestation varies by host immunity development, hypersensitivity to mite antigens, stress of confinement, and environmental conditions. Psoroptic mites are usually more active in winter. In the warmer months, lesions are often not obvious and can be missed.

In sheep, infestations tend to occur in the more densely wool-covered body regions, with initial lesions usually appearing on the backs and sides.

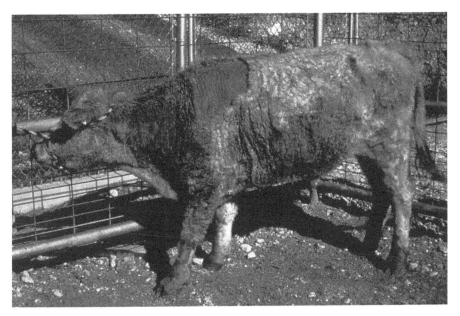

Figure 7.8
Cattle scabies infestation. (Photo reproduced with permission by U.S. Department of Agriculture, Agricultural Research Service [USDA-ARS].)

Noticeable crusting and scab formation will occur along with inflammation and wool and skin damage. The severe pruritis often associated with psoroptic mite infestation induces self-inflicted trauma resulting in extensive wool loss and skin injury due to animals licking, biting, and rubbing infested areas. In heavily infested, untreated animals, massive lesions can develop over the body causing extensive wool loss and reduced weight gain. Psoroptic scab in sheep can affect animals of all ages but is often more severe in young lambs. Death can sometimes occur in extremely severely infested sheep and lambs.

In cattle, *P. ovis* causes similar hair loss and lesion development as seen in sheep. The severity of infestations in cattle varies from being mild to extensively covering the entire body surface (Figure 7.8). Individual animal response will vary by the ability of an animal to develop immunity to mite antigens. In animals not previously exposed to these mites, mite populations often develop rapidly. In previously exposed animals that have developed an acquired immunity, there is often reduced growth rates and lower fecundity in mite populations. Cattle can become hypersensitive to mite antigens, however, and develop severe clinical signs of mite infestation. Stressed cattle kept in confinement and in close contact and cattle exposed to extremely cold weather are often more affected by mite infestations, with death sometimes occurring.

Hosts: Sheep, cattle, and sometimes, but rare, equines and other animals

Description: Adult female *P. ovis* mites are about 0.75 mm long. Males are somewhat smaller. In females, the legs are approximately the same length, but in males the fourth pair of legs is very short. Also, distinguishable with *P. ovis* is the trumpet-shaped pulvilli attached to three-jointed pretarsi. In most other mites, the pretarsi are unjointed.

Biology/Behavior: The complete lifecycle occurs on the host. Adult females produce from one to three eggs per day over about a 40- to 60-day period. There are two nymphal stages that follow the larval stage: protonymphs and deutonymphs. The lifecycle from egg to adult takes around 14 days. The mites can spread rapidly between animals by direct contact between animals or indirectly by infested animals rubbing against fence posts or other objects. Detached mite-laden scabby lesions then make contact with other animals.

Sarcoptes scabiei *(De Geer) (Order Astigmata; Family Sarcoptidae) (Figure 7.9)*

Common names: Sarcoptic mange, scabies mite, itch mite, hog mange mite, others

Geographic Distribution: Worldwide

Veterinary Importance: Although classified as one species, there are several host-specific varieties of *Sarcoptes scabiei*. This mite occurs on a wide range of wild and domestic animals, as well as on humans. However, populations that are prevalent on one host species do not readily infest another host species. Commonly known as sarcoptic mange mites, they affect a variety of domestic animals including dogs, swine, sheep, goats, cattle, and horses. It is relatively rare in cats. In humans, these mites are called human scabies or human itch mites.

Sarcoptes scabiei damage their host by adult female mites burrowing under the skin surface in the epidermis. This burrowing activity and feeding upon dissolved skin tissues cause varying degrees of dermatitis resulting in pruritis, hair loss, and dermal crusty lesions. Initial lesions can occur anywhere on the host body but are usually more common where the hair is more sparse, often on the head area. The infestation can then spread quickly to other areas of the body. As it progresses, the skin in affected areas becomes thickened, crusted scabs form that tend to crack and ooze exudates, and extensive hair loss can occur (Figure 7.10). Scratching and rubbing by the host causes further damage, and secondary infections can be an issue. Severely infested animals experience weight loss, reduced feed efficiency, reproductive losses, reduced milk production, lethargy, and sometimes death. In farm animals, infestations are more prevalent in the winter months due to environment stress and animal crowding.

Hosts: Humans and many wild and domestic animals

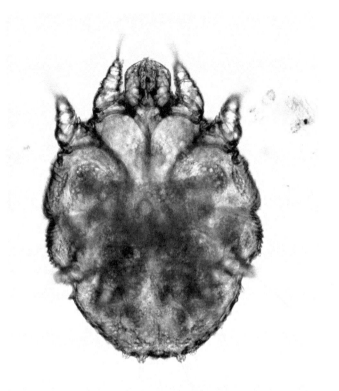

Figure 7.9
Sarcoptes scabiei. *(Photo reproduced with permission by Michael Dryden, Kansas State University.)*

Description: Adult *Sarcoptes scabiei* have a rounded body with small pointed triangular dorsal spines and short legs. Adult females are about 0.4 mm long and males about 0.2 mm long.

Biology/Behavior: Adult female *Sarcoptes* mites make an initial pit burrow in the skin of their host. There the adult female awaits a male to mate. Upon mating, the fertilized female moves on the skin and finds a site to dig a permanent burrow. Under the skin, she excavates a horizontal burrow within the stratum corneum that can be up to 1 cm long or longer. In the burrow, the female mite will deposit two to three eggs per day over about a 2-month period. The eggs hatch in 3 to 4 days. Emerged larvae crawl out of the burrow and onto the skin surface. After 2 to 3 days, the larvae molt into nymphs. Two nymphal molts occur: protonymphs and deutonymphs. The larvae and nymphal stages often make temporary burrows to feed and seek shelter. After 3 to 4 days in these nymphal stages, they then molt into adults. Upon emergence, adult mites move about on the skin, up to 2.5 cm/min feeding

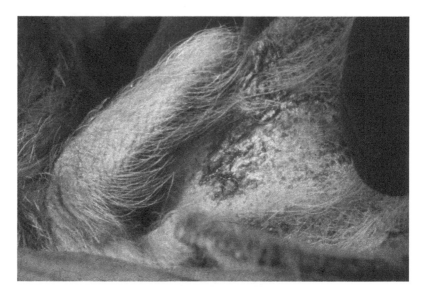

Figure 7.10
Sarcoptic hog mange in ear.

and burrowing on the skin surface. The complete lifecycle takes from 10 to 20 days.

Chorioptic Mange Mite (Order Astigmata; Family Psoroptidae) (Figure 7.11)

Species Name: *Chorioptes bovis* (Gerlach)

Geographic Distribution: Worldwide in temperate regions

Veterinary Importance: The chorioptic mange mite is most often a problem in domestic ungulates, primarily cattle, sheep, goats, and horses. Infestations primarily occur on the lower extremities of the legs and feet and the underside on the abdomen, udder, scrotum, and base of the tail. These mites do not burrow into the skin. However, their feeding activity on sloughed epidermal tissue often causes irritation with resultant crusty, pruritic, scabby lesions forming that warrant treatment. These clinical symptoms normally show up in heavier infestations. In male animals, especially rams, infestations on the scrotum can lead to temporary infertility. These mites are more problematic in winter months.

Hosts: Various ungulate animals

Description: Adult mites are similar in appearance to *Psoproptes ovis* but are smaller, being about 0.3 mm long. They do not have jointed pretarsi, however, as seen with *P. ovis*.

Figure 7.11
Chorioptic mange mite, Chorioptes bovis. *(Photo reproduced with permission by Marcelo de Campos Pereira, University of São Paulo, Brazil.)*

Biology/Behavior: Adult female chorioptic mange mites deposit one egg per day, attaching it to the host skin. They produce from 14 to 20 eggs during their lifetime of about 2 weeks. Because infestations occur in specific areas on a host, eggs are often clustered together. The eggs hatch in approximately 4 days. Larval and nymphal stages take from 10 to 15 days to complete. The total lifecycle from egg to adult takes about 3 weeks.

Sheep and Cattle Itch Mites (Order Prostigmata; Family Psorergatidae) (Figure 7.12)

Species Names:

Psorobia ovis (Wormersley)—sheep itch mite
P. bos (Johnston)—cattle itch mite

Geographic Distribution: *Psorobia ovis* has been found in Australia, New Zealand, Africa, South America, and the United States. *Psorobia bos* has been found in the United States, South Africa, and Great Britain.

Veterinary Importance: These two *Psorobia* species were formerly known in the genus *Psorergates. Psorobia bos* generally causes only mild symptoms in cattle. Some reports indicate observing some hair loss, slight skin thickening, and scaling on host animals. Infestations, when they occur, are often seen on the thorax, back, abdomen, and upper parts of the legs. Slight weight loss may occur in infested animals.

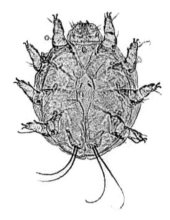

Figure 7.12
Sheep itch mite, Psorobia ovis.

The sheep itch mite, *P. ovis*, is considered of more significant veterinary importance. Severe infestations have been reported causing significant economic problems in sheep. Initial signs of infestations include small, pale areas of the wool on the shoulders, thighs, and/or flanks. The infestation will gradually extend over the rest of the body. Heavy infestations can cause severe irritation with sheep rubbing and biting at the infested areas causing further skin and wool damage. The wool becomes dry, stringy and/or matted, and discolored, turning a yellow-orange color. Also, scurf builds up on the skin where the mites are feeding. Infestations spread slowly over infested animals and through a flock and are most prominent in the winter months.

Hosts: *Psorobia bos* occurs on cattle and *P. ovis* on all breeds of sheep.

Description: The body of each of these mites is nearly circular with the legs arranged equally spaced around the body. Adults are about 0.2 mm long.

Biology/Behavior: The entire lifecycle is completed on the host and takes about 2 to 3 weeks. The mobile adults are very sensitive to desiccation and can survive for only 1 to 2 days off the host. These mites have three nymphal stages: protonymph, deutonymph, and tritonymph.

Demodex *Mites (Order Prostigmata; Family Demodicidae) (Figure 7.13)*

Species Names

Demodex bovis (Stiles)—cattle follicle mite
D. canis (Leydig)—dog follicle mite

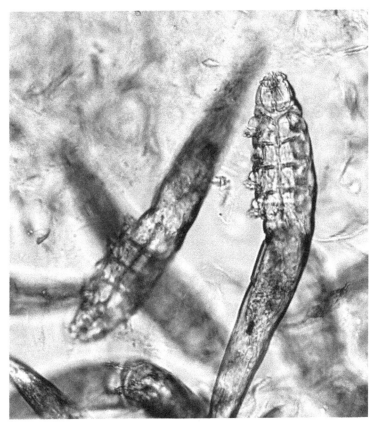

Figure 7.13
Dog follicle mite, Demodex canis. *(Photo reproduced with permission by Michael Dryden, Kansas State University.)*

   D. caprae (Railliet)—goat follicle mite
   D. cati (Megnin)—cat follicle mite
   D. equi (Railliet)—horse follicle mite
   D. folliculorum (Simon)—follicle mite
   D. phylloides (Csoker)—hog follicle mite

Geographic Distribution: Worldwide in association with host animal species.

Veterinary Importance*:* *Demodex* spp. are found on a wide range of wild and domestic animals and humans. Each species is very host specific. For the most part, these mites are nonpathogenic and are very commonly found on their hosts. They are typically found in hair follicles or dermal glands. Clinical demodectic mange is often most severe in dogs. However, livestock (such as cattle, goats, sheep, and swine), companion animals (such as equine and cats), and wild animals (such as foxes and other canids), and rodents can have clinical symptoms.

Host injury is caused by mites puncturing epithelial cells in hair follicles and glands with their chelicerae to feed. Host response is normally insignificant. In some cases, infested hair follicles and glands swell and become blocked resulting in the formation of dermal papules and nodules and often a scaly dermatitis develops. Hair loss can occur, and secondary bacterial infections can cause inflammation, pruritis, and pustule formation. Clinical lesions are often seen first on the face and head, and eventually spread to other parts of the body.

Demodectic mange in dogs can sometimes be severe, most often being seen in younger animals that are physiologically more susceptible. Initial clinical infestations in dogs are often seen around eyes and corners of the mouth. Most cases resolve themselves without treatment and the disease does not spread. However, in genetically predisposed or immunologically depressed animals, the disease can spread to other parts of the body causing extensive pustular demodicosis and hair loss. The skin can become tender, cracked, and sore, and may easily bleed. Because of the redness of the skin that results, this is often called red mange and usually has a distinctive rancid odor. The pathogenesis of demodicosis in dogs is believed to involve host immunosuppression caused by dermal staphylococcal infections.

It is thought that in the immune response by the host to this infection, that the host's cutaneous cell-mediated immunity is reduced, allowing the proliferation of the *Demodex* mite population. Most dogs become infected as new pups from their mother.

With the other *Demodex* species, varying degrees of demodicosis can occur in host species, but generally not near to the problematic degree as can occur in dogs. Occasionally, dermal demodectic mange will develop in infected host animals accompanied by patchy hair loss. In swine, the appearance takes on a rash appearance with numerous pustules. In cats, demodicosis is believed to become an issue in immunosuppressed cats resulting from diseases such as leukemia or diabetes mellitus. In cattle, *D. bovis* can sometimes produce enlarged dermal papules or cysts. Skin damage that results from the rupture of these papules can cause damage to the hide and economic losses due to reduced hide quality.

Hosts: Hosts are specific as to species and common names.

Description: *Demodex* mites are very small (0.1 to 0.4 mm long) with elongated tapered bodies. They have very short, stout, three-segmented legs. They lack setae on their legs and body. They have a pair of tiny, needle-like chelicerae adapted for piercing into host cells to feed.

Biology/Behavior: These mites live their entire lives in hair follicles and associated ducts and glands. They cannot survive off their host. They have the typical life stages of egg, larva, protonymph, deutonymph, and adult, all developing in the dermal hair follicles or glands. The entire lifecycle takes from 3 to 4 weeks.

Rabbit Ear Mite (Order Astigmata; Family Psoroptidae) (Figure 7.14)

Species Name: *Psoroptes cuniculi* (Delafond)

Geogaphic Distribution: Worldwide

Veterinary Importance: This mite is found primarily in rabbits and sometimes in commercial and scientific rabbit colonies. Infestations are usually localized in the ears, causing ear mange or psoroptic otocariasis. Scab formation in the external ear canal often takes on a honeycomb appearance. Heavy infestations may cause extensive scab development, blocking the auditory canal. Affected rabbits may appear lethargic or be seen digging frantically at the ears. Discoloration may be seen at the exterior ear openings. If mites penetrate into the inner ear an infection may develop, causing rabbits to hold their heads down to one side or another from the trauma caused. Occasionally, infestations may also spread to other body regions.

Hosts: Rabbits

Figure 7.14
Rabbit ear mite, Psoroptes cuniculi.

Description: This mite is morphologically similar to *P. ovis*.

Biology/Behavior: Other than being seen primarily in the ears, the lifecycle is similar to that of *P. ovis*.

Ear Mite (Order Astigmata; Family Psoroptidae) (Figure 7.15)

Species Name: *Otodectes cynotis* (Hering)

Geographic Distribution: Worldwide

Veterinary Importance: This is a very common mite found in cats and dogs and other carnivores. It infests the host deep in the external ear canal and causes otodectic mange. In heavy infestations, it may also be found infesting other body regions including the head, back, tail, and feet. It does not burrow

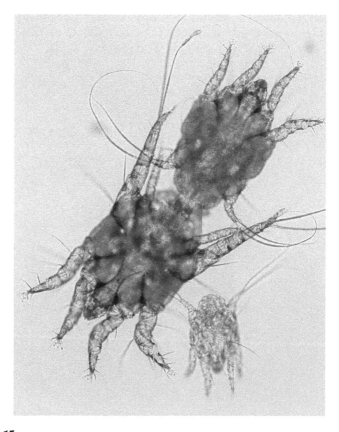

Figure 7.15
Ear mite, Otodectes cynotis. *(Photo reproduced with permission by Michael Dryden, Kansas State University.)*

into the skin but feeds externally, piercing the skin to feed on blood, serum, and lymph, and in the ears on aural exudates.

Most all cats and dogs usually possess small populations of this mite with no clinical signs. Allergic hypersensitivity by the host animal results in clinical otodectic mange. In sensitive animals, a crust may develop over the feeding mites. In the ears, accumulation of moist, brown to black exudates develops. Inflammation and pruritis may result. Intense irritation causes infested animals to scratch their ears and shake their heads. Secondary bacterial infection may occur. In severe cases, fever and depression may occur. Heavily infested dogs may act violently to infection, making them behave erratically. Transmission occurs by direct contact between animals and from mothers to their offspring.

Hosts: Dogs, cats, and other carnivores

Description: This mite closely resembles the *Psoroptes* sp. mite. It differs by having short, unjointed pretarsi with cup-shaped sucker-like pulvilli. Adult females have greatly reduced hind legs terminating in two, long, whiplike setae.

Biology/Behavior: The lifecycle is similar to that of *Psoroptes* sp. and is completed in about 3 weeks.

Notoedric Cat Mite (Order Astigmata; Family Sarcoptidae) (Figure 7.16)

Species Name: *Notoedres cati* (Hering)

Geographic Distribution: Worldwide

Veterinary Importance: This mite primarily infests wild and domestic cats, and occasionally rabbits and canids, including dogs. Lesions usually appear first on the ears, neck, face, and shoulders. As the infestation spreads, lesions sometimes will then be seen on the abdomen, legs, and genital area. In infested areas, crust and scab formation develops, similar to that of *S. scabiei*. The resulting dermatitis causes intense pruritis, erythema, crusty scaling of the skin, and hair loss. The skin becomes thickened and wrinkled. Infested animals will continuously scratch. If untreated, infested animals can become severely debilitated and infestation could lead to death.

Hosts: Wild and domestic cats and occasionally rabbits, dogs, and other canids

Description: This mite is similar in appearance to *S. scabiei*, but it lacks dorsal spines and has the anal opening located dorsally. They are also smaller than *S. scabiei*, being about 0.15 to 0.225 mm in length.

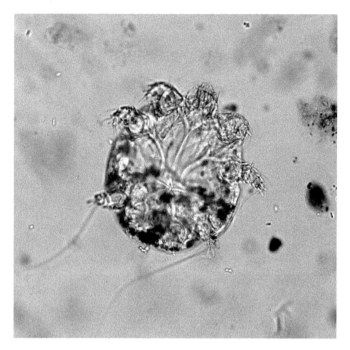

Figure 7.16
Notoedric cat mite, Notoedres cati. *(Photo reproduced with permission by Michael Dryden, Kansas State University.)*

Biology/Behavior: The lifecycle of this mite is very similar to that of *S. scabiei.* Adult females deposit eggs as they burrow in the skin. The eggs hatch in 3 to 4 days, and total development from egg to adult takes from 6 to 10 days.

Scaly-Leg Mite (Order Astigmata; Family Knemidokoptidae) (Figure 7.17)

Species Name: *Knemidocoptes mutans* (Robin and Lanquetin)

Geographic Distribution: Worldwide in areas associated with poultry.

Veterinary Importance: This mite burrows into the skin of the legs and feet of poultry, causing inflammation and formation of vesicles and white, scaly encrustations (Figure 7.18). Heavy infestations become very irritating and may cause lameness, deformity of legs and feet, and occasionally loss of claws. This condition is known as scaly-leg. The skin of the comb and wattle may also be affected. In severe infestations, birds stop feeding and ultimately die. Scaly-leg is more common in birds raised on the ground or in roosts, not in cages. It is very contagious between birds.

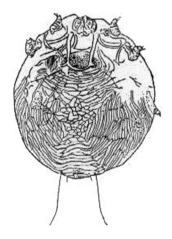

Figure 7.17
Scaly-leg mite, Knemidocoptes mutans.

Figure 7.18
Scaly-leg mite infestation. (Photo reproduced with permission by Virginia Martin,
www.feathersite.com.)

Hosts: Poultry and other birds

Description: Adult females are round in appearance and around 0.4 mm
long. They have short, stubby legs, and the dorsal surface is covered with light
striations and scale-like patterns. The body lacks spines.

Biology/Behavior: Adult females give birth to larvae, as opposed to eggs,
directly in the skin burrows. There are two nymphal molts. The entire lifecycle
takes from 2 to 3 weeks.

Depluming Mite (Order Astigmata; Family Knemidocoptidae)

Species Name: *Knemidocoptes gallinae* (Railliet)

Geographic Distribution: Worldwide

Veterinary Importance: The depluming mite burrows into the base of feather shafts. It is most frequently found affecting birds on the head, back, neck, abdomen, around the vent area, and upper legs. Infestations cause intense irritation. Scratching and pecking by birds results in noticeable feather loss. Feathers easily break off at the base. This depicts the common name of the depluming mite.

Hosts: Poultry and other birds, including pet species

Description: This mite is similar in appearance to *K. mutans*. The main difference is that the dorsal striations are continuous and not broken as in *K. mutans*.

Biology/Behavior: This mite has a similar lifecycle as *K. mutans* except for the body region where it primarily infests as discussed above.

INTERNAL FEEDING MITES

Cattle Ear Mite (Order Mesostigmata; Family Halarachnidae)

Species Name: *Raillietia auris* (Leidy)

Geographic Distribution: Found in North America, Europe, Western Asia, and Australia

Veterinary Importance: This mite is considered relatively harmless. It feeds primarily on skin detritus and wax in the ear canals of cattle. It can cause blockage of the auditory canal. In severe cases, ear canal inflammation can occur as well as pus formation and ulcerated lesions and hemorrhaging. Animals may experience hearing loss.

Hosts: Dairy and beef cattle

Description: The cattle ear mite is a small mite about 1 mm long and oval in shape.

Biology/Behavior: Adult mites crawl in the ear canal to feed and mate. The female then oviposits eggs, and larvae emerge in the ear canal. After feeding, the larvae leave the host and molt into protonymphs followed by deutonymphs. Deutonymphs then molt into adults. The complete lifecycle takes about 8 days.

Fowl Cyst Mite (Order Astigmata; Family Laminosioptidae) (Figure 7.19)

Species Name: *Laminosioptes cysticola* (Vizioli)

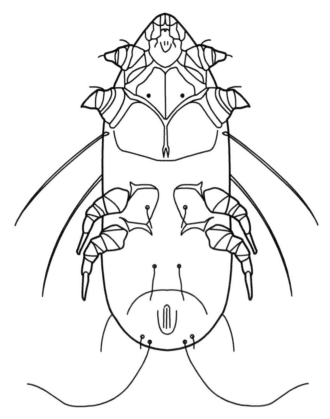

Figure 7.19
Fowl cyst mite, Laminosioptes cysticola. *(Illustration courtesy of Sarah Cox, Purdue University.)*

Geographic Distribution: Abundant in Europe and also found in the United States, South America, and Australia

Veterinary Importance: This mite is an internal parasite of poultry occurring in the subcutaneous tissue in the neck, breast, flanks, and vent. Small nodules form that become calcified. In heavy infestations, several hundred of these nodules may be present. These calcified nodules can reduce carcass value.

Description: This is a small mite, about 0.25 mm long. They have a smooth, elongated body with few setae.

Biology/Behavior: Very little is known about the lifecycle of this mite.

Airsac Mite (Order Astigmata; Family Cytoditidae) (Figure 7.20)

Species Name: *Cytodites nudus* (Vizioli)

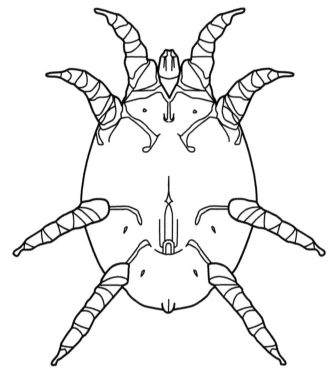

Figure 7.20
Airsac mite, Cytodites nudus. *(Illustration courtesy of Sarah Cox, Purdue University.)*

Geographic Distribution: North America, South America, New Zealand, Australia, and India

Veterinary Importance: This mite infests the bronchi, lungs, airsacs, and bone cavities of birds. Small infestations usually have no clinical effect on their hosts. Heavier infestations may cause coughing and accumulation of mucous in the trachea and bronchi. Emaciation, peritonitis, pneumonia, and obstruction of air passages can result in weakness and weight loss.

Hosts: Domestic and other birds, including pheasants, pigeons, canaries, and ruffed grouse

Description: These mites are small, about 0.6 mm long, and oval in shape. They have a smooth cuticle with only a few short setae. The gnathosoma is reduced with the chelicerae absent and the palps fused into a sucking organ.

Biology/Behavior: Little is known of the lifecycle of this mite. It is thought that eggs are laid in the lower air passages. Infestations are probably spread through coughing by the infested host bird.

BENEFICIAL MITES

House Fly Mite (Order Mesostigmata; Family Macrochelidae)

Species Name: *Macrocheles muscaedomesticae* (Scapoli)

Geographic Distribution: Worldwide

Veterinary Importance: This mite is predaceous on house flies, *Musca domestica*, feeding on fly eggs and first-instar larvae. Individual mites are capable of preying on 10 to 20 immature flies per day. They are most frequently found in accumulated manure under caged chickens, especially at the top of the manure cones where flies normally deposit their eggs. They have been considered as a biological control agent of house flies and, as such, play a role in the integrated pest management approach to fly control in confined livestock and poultry production systems.

Hosts: These mites are predaceous on *Musca domestica* and probably on other muscoid species.

Description: Adult mites are reddish-brown in color, heavily sclerotized, and about 1 to 1.5 mm long.

Biology/Behavior: The life stages of this mite include egg, larva, protonymph, deutonymph, and adult. The mite's complete lifecycle takes only about 2 days. Adult females live for about 24 days and deposit most of their eggs in the first 2 weeks of their adult lives. They normally disperse to new habitats by phorectic transport of adult mites on flies.

Fuscuropoda vegetans (De Geer) (Order Mesostigmata; Family Uropodidae)

Geographic Distribution: Widespread in North America

Veterinary Importance: This mite can be found in the same manure habitats as *Macrocheles muscaedomesticae*, especially in accumulated manure under caged chickens. These mites prey mostly on first-instar fly larvae, but not eggs. They are unable to pierce the chorion of fly eggs. Dissemination is by phorectic attachment onto dung beetles.

Hosts: Predaceous on first-instar house fly larvae and probably other muscoid species. It will also prey on nematodes.

Description: This mite is similar in size to *M. muscaedomesticae*, slightly oval, heavily sclerotized, and reddish-brown in color.

163

Biology/Behavior: The lifecycle of this mite takes from 30 to 40 days to complete, and it can become the most abundant mite in accumulated manure in 8 to 12 weeks.

CHAPTER 8

TICKS

TABLE OF CONTENTS

Ticks are a destructive group of blood-sucking parasitic arthropods found throughout the world. Although there are only about 800 species of ticks known, they are a very important arthropod group of veterinary concern. They are morphologically and physiologically capable of surviving long periods between blood meals, surviving for up to several years. When they periodically feed, they take large blood meals. Tick bites may cause direct damage to domestic animals through blood loss, irritation, inflammation, and development of hypersensitivity resulting in anemia and reduced productivity. Some species of ticks may cause toxicosis and paralysis from their salivary secretions. Ticks are also capable of transmitting a variety of pathogens, including protozoan, viral, bacterial, rickettsial, and fungal pathogens. Costs associated with tick damage to animal production and well-being and costs associated with animal treatment and premise treatments for tick control amount to millions of dollars annually.

TAXONOMY/MORPHOLOGY

Ticks are in the class Arachnida and subclass Acari and are closely related to mites. Within the order Parasitiformes, ticks are represented by three families: Ixodidae, Argasidae, and Nuttalliellidae. Nuttalliellidae contains a single species that is not of veterinary importance. Ixodidae, the hard ticks, contains most of the species of veterinary importance. Argasidae, the soft ticks, contains a few species of veterinary importance (Table 8.1).

All ticks have two main body regions (Figure 8.1). These include the anterior gnathosoma (or capitulum) bearing the mouthparts and the posterior idiosoma bearing the legs and unsegmented abdomen. The gnathosoma consists of the basis capituli with a pair of four-segmented palps arising laterally and cheliceral sheaths housing the two chelicerae. The distal end of the chelicerae is scleritized in a tooth-like pattern used to cut into the host's skin. Under the chelicerae is the hypostome that is inserted into the host. The idiosoma bears the legs. Larval ticks have six legs, and nymphs and adults have eight legs. The legs each have six segments, normally ending with paired tarsal claws and pad-like pulvillus. On the tarsi of the first pair of legs is a cavity known as Haller's organ. This houses numerous chemoreceptor setae and functions as an olfactory organ.

Ixodidae

Known as hard ticks, ixodid ticks show sexual dimorphism. Adult females possess a hard cuticular scutum restricted to the anterior dorsal surface behind the

166

Table 8.1
Ticks of Veterinary Importance

Species	Common Name
Family Ixodidae	
Amblyomma americanum (Linnaeus)	Lone star tick
A. cajennense (Fabricius)	Cayenne tick
A. maculatum (Koch)	Gulf Coast tick
Dermacentor albipictus (Packard)	Winter tick
D. andersoni (Stiles)	Rocky Mountain wood tick
D. nitens (Neumann)	Tropical horse tick
D. occidentalis (Marx)	Pacific Coast tick
D. variabilis (Say)	American dog tick
Ixodes scapularis (Say)	Black-legged tick
Rhipicephalus (Boophilus) annulatus (Say)	Cattle tick (cattle fever tick)
R. (Boophilus) microplus (Canestrini)	Southern cattle tick (Tropical cattle tick)
R. sanguineus (Lareille)	Brown dog tick
Family Argasidae	
Argas miniatus (Koch)	Fowl ticks
A. radiatus (Raillet)	
A. persicus (Oken)	
A. sanchezi (Duges)	
Ornithodoros coriaceus (Koch)	Pajaroello tick
Otobius megnini (Duges)	Spinose ear tick

capitulum. This allows for the female to take a large blood meal and expand significantly in size. In males, the entire dorsal surface behind the capitulum is almost completely covered by the scutum. The scutum in both females and males is often ornate with species-specific color patterns. Adult males are generally smaller than females. Females stay attached to their host for several days and take in one large blood meal. When they detach, they will deposit one batch of eggs and then die. Males may feed several times, or not at all, only ingesting small quantities of blood at a time. In larvae and nymphs, the dorsal scutum is also restricted anteriorly, allowing them to engorge as they feed. When eyes are present in ixodid ticks, they are found on the lateral margin of the scutum. Also distinguishable in ixodid ticks is that the capitulum extends anteriorly from the body and is visible from above.

Argasidae

Known as soft ticks, the body of argasid ticks is leathery, unsclerotized, and has a textured surface. The surface is characteristically marked with grooves

Ixodes scapularis
(Black-legged tick)
Note: Drawing is not to scale
Actual Size:

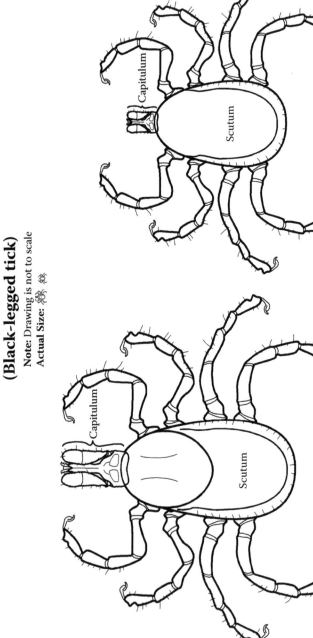

Adult Male

Adult Female

Figure 8.1
Tick diagram, Ixodes scapularis. *(Illustration courtesy of Scott Charlesworth, Purdue University.)*

and folds. The integument is inornate, and there is little sexual dimorphism. The capitulum of the larvae extends out from the anterior body margin, but in nymphs and adults it is ventrally located and generally not visible dorsally. When present, the eyes are positioned in lateral folds above the legs.

LIFE HISTORY/BIOLOGY

The lifecycle of ticks includes four stages: egg, six-legged larva, eight-legged nymph, and eight-legged adult (Figure 8.2). In ixodid ticks, there is one nymphal instar, and in argasid ticks there are from two to seven nymphal instars. Larvae, nymphs, and adults of most species all feed on blood. In ixodid ticks, all three of these stages, as well as larvae and nymphs of some argasid ticks, feed for long periods of time (from 2 to 14 days). Most adult and many immature argasid ticks will feed for much shorter intervals from a few minutes to a few hours.

In the feeding process, the tick cuts the skin of the host with their chelicerae. They then insert the hypostome with the palps spreading out on the host skin surface. Ticks with short mouthparts use salivary gland secretions to cement the mouthparts to the host. Anticoagulants and other substances are also secreted to aid in feeding success.

Lifecycles vary greatly between different tick species. They can be categorized according to the number of hosts fed upon and off-host periods of different life stages. With ixodid ticks, lifecycles can be categorized by the number of hosts utilized: one-host, two-host, or three-host ticks.

One-Host Ticks

With these tick species (e.g., *Rhipicephalus (Boophius) spp.*, *Dermacentor albipictus*, *D. nitans*), unfed larvae attach to the host and remain on the single host through three separate blood feedings of the larval, nymphal, and adult stages. During any of these stages, they may reattach on the host several times before complete engorgement. Adult males will mate with partially engorged females while they are attached. Replete adult females will drop to the ground to oviposit eggs. Depending on the species and such factors as environmental conditions and host density, one-host ticks may produce two or more generations per year. The spinose ear tick, *Otobius megnini*, is considered a one-host tick, but it differs in that engorged second-instar nymphs, after feeding, drop to the ground. Upon adult emergence, males and females mate and oviposition occurs on the ground. Larvae will then seek a host and attach to the inner ears of cattle. With one-host ticks, they are usually somewhat host specific.

Two-Host Ticks

In these species (e.g., *Rhipicephalus evertsi*), larvae and nymphs feed on the same host with engorged nymphs dropping to the ground. Emerged adults then seek another host to feed upon.

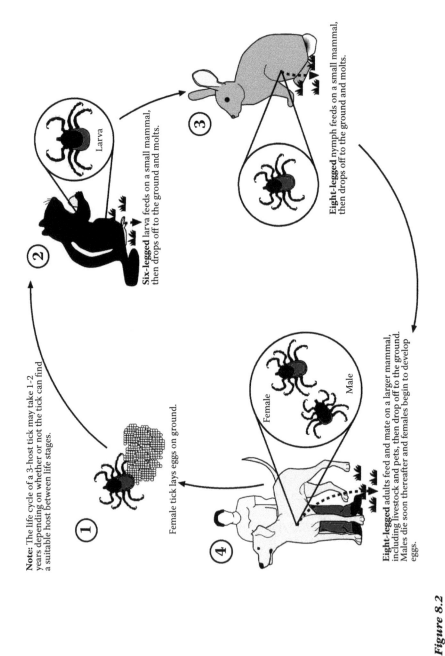

Note: The life cycle of a 3-host tick may take 1-2 years depending on whether or not the tick can find a suitable host between life stages.

Female tick lays eggs on ground.

Six-legged larva feeds on a small mammal, then drops off to the ground and molts.

Larva

Eight-legged nymph feeds on a small mammal, then drops off to the ground and molts.

Female

Male

Eight-legged adults feed and mate on a larger mammal, including livestock and pets, then drop off to the ground. Males die soon thereafter and females begin to develop eggs.

Figure 8.2
Generalized three-host tick lifecyle. (Illustration courtesy of Scott Charlesworth, Purdue University.)

Three-Host Ticks

In these ticks (e.g., *Dermacentor variablis*, *Amblyomma americanun*, others), a separate host is used for each feeding stage. Upon the completion of each blood meal, engorged larvae and nymphs drop to the ground to molt. Adult females upon repletion drop to the ground to oviposit eggs. Mating generally occurs on the host except in some *Ixodes* sp. Three-host ticks are generally less host specific than one-host ticks. With some species, immature stages may feed upon unrelated host species (e.g., birds, rodents). Adults tend to feed on larger animal hosts (e.g., deer, cattle, dogs). Other species, such as *Amblyomma americanum,* have less host size preference with each feeding stage. With three-host ticks, there is often seasonal activity associated with each stage. There is usually one generation per year but can extend to 2 years, depending on species, environmental factors, and host availability.

Most argasid tick species have multihost lifecycles. Separate hosts are generally used for larvae, each of the nymphal instars (from two to seven instars), and adults. Adult argasids may actually feed several times with periods of time spent away from the host between feedings.

Host-seeking behavior in ticks varies somewhat. Most tick species are relatively immobile and wait for hosts to find them. They will climb onto vegetation to await their host. These ticks detect host presence by odor (e.g., carbon dioxide, skin extracts), radiant heat, visual presence, and vibrations. They extend their forelegs outward and cling to the hair, feathers, skin, or clothing of the passing host. They then move about on the host to find a suitable attachment site. Some tick species, when they detect host presence, will emerge from ground cover and move to their host.

IXODID TICKS OF VETERINARY IMPORTANCE

Lone Star Tick (Order Acari; Family Ixodidae) (Figure 8.3)

Species Name: *Amblyomma americanum* (Linnaeus)

Geographic Distribution: From the Atlantic Coast from New York to Florida, west to Missouri, Oklahoma, and Texas

Veterinary Importance: This is a common tick species found within its distribution range. Larvae, nymphs, and adults will readily attack a variety of animals from small rodents and ground-nesting birds to larger animals, both domestic and wild, and humans. Lone star ticks are most frequently found in wooded areas supporting adequate populations of wildlife hosts. They are often abundant in areas with large populations of deer, which serve as the primary host for adult ticks. On livestock and companion animals, lone star

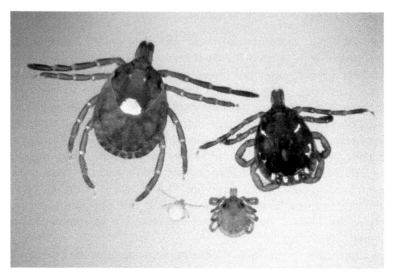

Figure 8.3
Lone star tick adults, nymph, larva, Amblyomma americanum. *(Photo reproduced with permission by Jim Kalisch, University of Nebraska.)*

ticks can be of significant importance. In heavily infested areas, host animals have been observed to harbor hundreds to thousands of lone star ticks at any given time. In cattle, lone star tick infestations will be found predominately around the tail head, escutcheon, udder, scrotum, ancillaries, and dewlap. The mouthparts of lone star ticks are capable of deep penetration, causing painful reactions in the host. Host reaction to tick attachment results in inflammation, maturating sores, and tenderness. Secondary infection will often occur at the attachment site. Heavily infested animals show reduced performance, weight loss, reduced milk production, and reproductive complications. Animals also become more susceptible to other infections. Significant mortality in white-tailed deer fawns has been associated with heavy infestations of lone star tick. Lone star ticks have been incriminated as possible vectors of *Ehrlichia*, the causative agent of ehrlichiosis.

Hosts: Found on a large variety of hosts including rodents, ground-inhabiting birds, deer, cattle, horses, sheep, dogs, and several other animal species, and humans.

Description: Adults are brown to tan in color and about 8.5 mm long unengorged. Adult females are readily distinguished by having a single silvery-white spot near the posterior of the dorsal scutum, while the scutum of adult males have scattered spots and markings around the dorsal margins. The mouthparts in this genus are noticeably longer than other genera of ixodid ticks.

Biology/Behavior: Adult females of this three-host tick are common in late spring and early summer. Larvae, often called *seed ticks,* are common in mid-June through July. Nymphs often have two activity peaks. The first peak comprising of overwintering nymphs occurs between April and June. The second peak, following larval activity, usually occurs in July and August. Variations will occur in peak activity due to geographic location, host availability, and seasonal climatic variations. There is usually one generation per year. They generally overwinter as unfed adults or nymphs under ground cover debris.

Gulf Coast Tick (Order Acari; Family Ixodidae) (Figure 8.4)

Species Name: *Amblyomma maculatum* (Koch)

Geographic Distribution: Found along the Gulf and Atlantic Coasts from South Carolina to Texas, northward to southeastern Kansas, and south to Mexico and Central and South America.

Figure 8.4 (see color insert following page 132)
(Top) Gulf Coast tick female, Amblyomma maculatum. *(Bottom) Gulf Coast tick male,* Amblyomma maculatum. *(Photos courtesy of James Gathany, Centers for Disease Control and Prevention.)*

Figure 8.5
"Gotch" ear caused by Gulf Coast ticks.

Veterinary Importance: The Gulf Coast tick is a serious pest of cattle within its distribution range. It will also infest other domestic animals (e.g., horses, sheep, swine, dogs) and wild animals (e.g., deer, coyotes). In range cattle, significant infestations can occur. Adult ticks prefer to attach and feed on cattle on the inside areas of the outer ears. Because of their long mouthparts and concentration in this small area of the host, intense irritation and host reaction often occurs. As many as 75 to 100 ticks per animal have been observed, confined to the ears. The infested ears become inflamed and sore with open, scabby lesions. As the ears thicken in response to tick feeding, they curl inward in a distinctive appearance known as *gotch ear* (Figure 8.5). The intense irritation results in less feed efficiency and grazing time with significant weight loss occurring. Infested ears develop secondary infections. Historically, the open sores caused by tick feeding have predisposed the ear to attack by the screwworm fly, *Cochliomyia hominvorax*. Pathological response reflective of altered blood chemistry has been documented in Gulf Coast tick–infested cattle. Gulf Coast ticks have also been found attaching onto the hump region of Brahman cattle and in the neck regions of all cattle breeds.

Hosts: Immature Gulf Coast ticks are found on small mammals and ground-inhabiting birds. Adults infest cattle, horses, sheep, swine, dogs, deer, coyotes, and other carnivores.

Description: Like the lone star tick, *A. americanum*, the Gulf Coast tick has long mouthparts. Adults are dark brown with the scutum of both females and males ornate with distinctive silver-white lines.

174

Biology/Behavior: Adult activity of this three-host tick varies in its geographic range. In coastal areas and into Texas, adults are active from early spring but reach peaks in August and September. Further north into northeastern Oklahoma and southeastern Kansas, adult tick activity will peak earlier in May and June. In the southeastern United States, peak activity has been observed in mid-July. Following adult peaks in activity, peaks in immatures will follow.

Upon repletion on the host, adult females drop to the ground to oviposit their egg mass of several thousand eggs. Larvae, and then nymphs, will feed primarily on small mammals and ground-inhabiting birds (e.g., meadowlarks, quail). Gulf Coast ticks favor grasslands and prairies where there are suitable hosts for both immatures and adults. As with *A. americanum*, there is usually one generation per year with unfed adults and nymphs overwintering.

Cayenne Tick (Order Acari; Family Ixodidae) (Figure 8.6)

Species Name: *Amblyomma cajennense* (Fabricius)

Geographic Distribution: Primarily confined in southern Texas in the United States but more prominent in Mexico and Central and South America

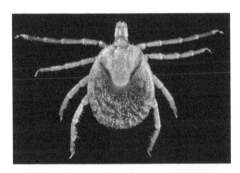

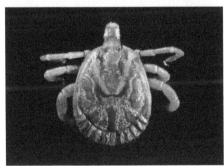

Figure 8.6
(Top) Cayenne tick female, Amblyomma cajennense. *(Bottom) Cayenne tick male,* Amblyomma cajennense. *(Photos reproduced with permission by James Gathany, Centers for Disease Control and Prevention.)*

175

Veterinary Importance: In areas where this tick is found, it can be of considerable importance as a cattle pest. Although it can be found year round, infestations are greatest on cattle from March to May. With long mouthparts as found in other *Amblyomma* ticks, *A. cajennense* can inflict painful wounds at their attachment sites. On cattle, they prefer attachment sites on the dewlap, axillaries, udder/scrotum, escutcheon, and tailhead area.

Hosts: Immatures prefer to feed on quail, wild turkeys, coyotes, raccoons, and a variety of other smaller wild animals. Adults will feed on deer, cattle, horses, sheep, goats, dogs, and a variety of other animal hosts.

Description: These ticks have long mouthparts. The scutum is ornate with a distinctive silver-white pattern on both adult female and male.

Biology/Behavior: Adults and nymphs of this three-host tick are most prevalent from March to May. They can develop year round with one or more generations per year depending on host availability.

Cattle Tick (Cattle Fever Tick) (Order Acari; Family Ixodidae)

Species Name: *Rhipicephalus (Boophilus) annulatus* (Say)

Geographic Distribution: This tick is found in subtropical and tropical areas. It is found in North Africa, western Asia and Middle East, Mediterranean, and parts of North America, Central America, and South America. With successful state and federal eradication efforts, *R. annulatus* has been confined to occasional infestations in South Texas and southern California within buffer quarantine zones along the United States–Mexico border.

Veterinary Importance: This tick is most notorious as an important vector of the protozoan parasite *Babesia bigemina*, the causative agent of Texas cattle fever. It can also transmit *Babesia bovis*. For this reason, eradication efforts were initiated and permanent quarantine zones were established along the Mexican borders. Before its eradication in the early 1900s, babesiosis cost the U.S. cattle industry an estimated current dollar equivalent of $3 billion in direct and indirect annual losses. Of historical significance, transmission of *Babesia bigemina* by *R. annulatus* to cattle, as discovered in 1893, was the first report of a disease to be transmitted by an arthropod. This tick can also transmit *Anaplasma marginale*, the causative agent of anaplasmosis in cattle.

In addition to its importance in disease transmission, heavy infestations of *R. annulatus* can result in decreased animal productivity and damaged hides.

Hosts: Primarily found on cattle but can occur on other animals such as horses, sheep, goats, dogs, cats, and deer.

Description: Taxonomically, this species was originally known as *Boophilus annulatus*. *Boophilus* has recently become a subgenus of the genus *Rhipicephalus*. *Boophilus* ticks have a distinctive hexagonal basis capitulum. The spiracular plates are rounded, and they have very short palps. The adults are small and have no festoons or ornamentation.

Biology/Behavior: This is a one-host tick with all feeding stages spent on a single host animal. Adult males mate with attached feeding females. Replete adult females detach from the host and fall to the ground where they oviposit their eggs and then die. Hatched larvae climb up on vegetation to find a host. Once on a host, larval ticks seek softer skin to attach to (e.g., inside the thigh, flanks, forelegs, or on the abdomen and brisket). Engorged larvae molt into nymphs that attach and feed, and they, in turn, molt into adults. Preferred sites for adults are the tailhead, escutcheon, udder/scrotum, neck, dewlap, and axillaries. The complete lifecycle can be completed in 3 to 4 weeks. There are normally from two to four generations per year, depending on climatic conditions and host availability.

Southern Cattle Tick (Tropical Cattle Tick) (Order Acari; Family Ixodidae) (Figure 8.7)

Species Name: *Rhipicephalus (Boophilus) microplus* (Canestrini)

Geographic Distribution: Found worldwide in subtropical and tropical areas. It is found in Australia, Asia, South Africa, North America, Central

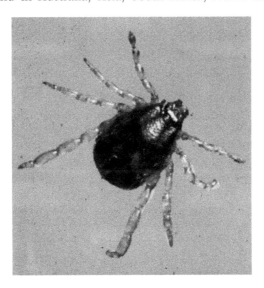

Figure 8.7
Southern cattle tick, Rhipicephalus (Boophilus) microplus. *(Photo courtesy of Mat Pound, USDA/ARS Bugwood.org.)*

America, South America, the Caribbean, and Mexico. With successful state and federal eradication efforts, *R. microplus* has been confined to occasional infestations in South Texas and southern California in buffer quarantine zones along the United States–Mexico border.

Veterinary Importance: This tick is considered to be the most important parasitic tick of livestock in the world because of its wide range of incidence. Heavy infestations result in loss of productivity and damaged hides. It is a vector of both *Babesia bigemina* and *B. bovis,* causing babesiosis in cattle. It can also transmit *Anaplasma marginale,* the causative agent of anaplasmosis. This tick along with *R. annulatus* are the targeted species for the eradication program as described above for *R. annulatus.*

Hosts: Primarily found on cattle but can occur on other animals including horses, other equine, sheep, goats, swine, dogs, and deer along with a variety of other wild animals.

Description: This tick is similar in appearance to *R. annulatus.* With *R. microplus,* adults have a short, straight capitulum. The legs are of a light-creamy color, and there is a wide space between the first pair of legs and the mouthparts.

Biology/Behavior: This one-host tick has a similar lifecycle to that of *R. annulatus.* In some tropical areas, *R. microplus* can have more than four generations per year.

American Dog Tick (Order Acari; Family Ixodidae) (Figure 8.8)

Species Name: *Dermacentor variabilis* (Say)

Geographic Distribution: Widely distributed in the eastern United States, and parts of the western United States, Canada, and Mexico.

Veterinary Importance: This tick is considered of relatively minor importance as a livestock pest. However, it has been demonstrated under controlled conditions to be able to transmit *Anaplasma marginale* in cattle, the causative agent of anaplasmosis. It is the most important vector of the causative agent of Rocky Mountain spotted fever (RMSF), *Rickettsia rickettsii,* in the eastern United States. Dogs infected with RMSF have shown clinical symptoms of fever and lethargy. American dog ticks are also known vectors of *Francisella tularensis,* the causative agent of tularemia. This tick has also occasionally been associated with tick paralysis in animals.

In heavy infestations, adult American dog ticks may be found over all parts of the host body, with preferred attachment sites being the neck, axillaries, groin, escutcheon, and under body surfaces.

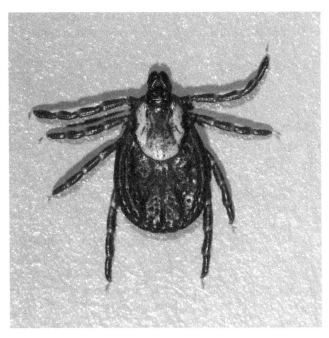

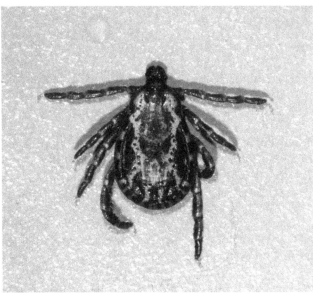

Figure 8.8
(Top) American dog tick adult female, Dermacentor variabilis. *(Bottom) American dog tick adult male,* Dermacentor variabilis. *(Photos reproduced with permission by Michael Dryden, Kansas State University.)*

Hosts: Larvae and nymphs feed primarily on small mammals, especially small rodents. Adults prefer larger animals, especially canines, but will feed on cattle, horses, a variety of medium to larger-sized wild animals, and humans.

Description: Adults are reddish-brown in color with an ornate, light grayish to white patterned scutum. Unfed females are about 4 mm long with males a little smaller. They have short mouthparts.

Biology/Behavior: The American dog tick is a three-host tick. Adult females attach and feed for 1 to 2 weeks. Males mate with attached, feeding females. Replete females drop to the ground and oviposit from 4000 to 6500 eggs and then die. Eggs hatch in about 4 weeks, and larvae seek a suitable host, usually a small rodent. Larvae feed from 4 to 5 days, drop to the ground, and molt into nymphs. Nymphs also prefer small rodents and feed for 5 to 6 days and then drop to the ground. In the southern range of this tick, all feeding stages may be found on their preferred hosts year round. In temperate regions, adults are found on their hosts from April to September, being most common in May and June. Under favorable conditions and an abundance of hosts, the lifecycle from egg to adult can occur in 3 months. However, in temperate regions, there is generally only one generation per year.

Rocky Mountain Wood Tick (Order Acari; Family Ixodidae) (Figure 8.9)

Species Name: *Dermacentor andersoni* (Stiles)

Geographic Distribution: Widely distributed in western North America from northern Arizona and New Mexico into Canada and eastward to North and South Dakota and western Nebraska.

Veterinary Importance: *Dermacentor andersoni* is the primary tick in the northwestern United States and Canada which causes tick paralysis in livestock (e.g., cattle, sheep). It has been shown to be a vector of *Anaplasma marginale*, the causative agent of bovine anaplasmosis. It is also the chief vector of *Rickettsia ricettsii,* the causative agent of Rocky Mountain spotted fever, in the western United States. Adult tick infestations, which are found on larger animals, are generally observed on their host around the head, neck, shoulders, groin, and escutcheon.

Hosts: Immatures feed primarily on small mammals (e.g., squirrels, rabbits, woodchucks, mice, shrews). Adults feed on larger hosts (e.g., cattle, horses, sheep, dogs, deer, elk, bear, coyotes).

Description: Adults are reddish-brown in color, about 4 to 6 mm long, and have short mouthparts. The female's scutum is almost solidly silver-white in

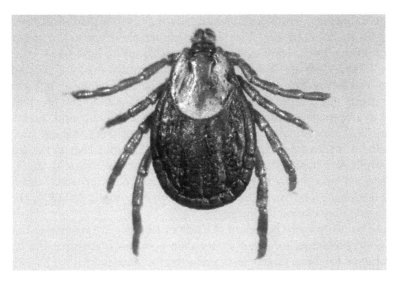

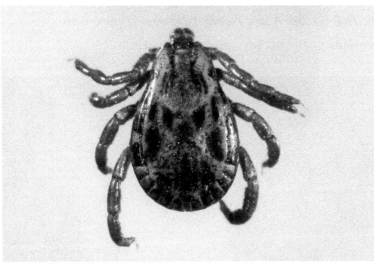

Figure 8.9
(Top) Rocky Mountain wood tick female, Dermacentor andersoni. *(Bottom) Rocky Mountain wood tick male,* Dermacentor andersoni. *(Photos reproduced with permission by Jim Kalisch. University of Nebraska.)*

color. The male scutum is ornate with silver-white markings. The scutum coloration of *D. andersoni* is brighter in appearance as compared to that of *D. variabilis.*

Biology/Behavior: The Rocky Mountain wood tick is a three-host tick. Males mate with attached, feeding females. Replete females drop to the ground and oviposit from 6000 to 6500 eggs over a 3-week period. Eggs hatch in about 4 weeks. Emerged larvae seek a suitable small mammal host and feed for 4 to 5 days. Fed larvae drop to the ground and molt into nymphs. Nymphs seek a host, feed, and drop to the ground and molt into adults. The lifecycle takes from 1 to 2 years to complete, depending on host availability and climatic conditions. Eggs are normally laid in early spring. Individuals that are successful in finding hosts feed as larvae in the spring and nymphs later in the summer. Successfully fed nymphs will molt and overwinter as adults to successfully complete a 1-year lifecycle. Nymphs that fail to feed overwinter and form a spring-feeding second-year lifecycle generation of nymphs. In colder climates, the lifecycle may extend to 3 years.

These ticks are more prevalent in areas of low, brushy vegetation that provides good protection for small mammalian hosts for immatures and has sufficient forage to attract larger hosts preferred for adult ticks.

Winter Tick (Order Acari; Family Ixodidae) (Figure 8.10)

Species Name: *Dermacentor albipictus* (Packard)

Figure 8.10
Engorged winter tick, Dermacentor albipictus. *(Photo reproduced with permission by Jim Kalisch, University of Nebraska.)*

Geographic Distribution: Widely distributed across the western United States and extending from northern Mexico into Canada.

Veterinary Importance: Heavy infestations of this tick on cattle and horses can cause significant loss of productivity with symptoms of lethargy, inappetence, lusterless hair coat, general debilitation, and edema, and even death from anemia. Infestations are generally found on the axillaries, dewlap, udder/scrotum, and escutcheon, but can be found anywhere on the body in heavy infestations.

Hosts: Deer, moose, elk, horses, other equine, and cattle

Description: The winter tick is similar in appearance and size to *D. variabilis* and *D. andersoni* with ornate scutum and short mouthparts.

Biology/Behavior: *Dermacentor albipictus* is a one-host tick that only infests animals from autumn until early spring (October to March/April) with greatest activity observed in December and January. The parasitic period on animals ranges from 4 to 8 weeks. Replete adult females drop to the ground and oviposit their eggs in the spring. Eggs hatch in 3 to 6 weeks. Emerged larvae remain clustered and inactive until cooler weather in late autumn when they then seek a host. Feeding and molting occur on the same host animal, and it takes about 4 to 6 weeks to complete feeding through the adult stage.

Tropical Horse Tick (Order Acari; Family Ixodidae)

Species Name: *Dermacentor nitens* (Neumann)

Geographic Distribution: Found only in southern Florida and southern Texas in the United States, but common in Mexico, Central America, the West Indies, and northern South America.

Veterinary Importance: Primarily found on equine, it is a serious pest of horses in tropical and subtropical areas. The preferred infestation site is in the external portion of the host's ears. In heavy infestations, ticks may be found spread out on other areas of the head, mane, and other body regions. Concentrated feeding in the ears it causes irritation, inflammation, and thickening of ear tissue. Horses become very sensitive while being handled. Accumulation of tick feces and cast exuviae from larvae and nymphs builds up in the ears producing an offensive odor. Heavy infestations in the ears predispose animals to attack by the screwworm, *Cochliomyia hominvorax*, in geographic areas where this parasitic fly exists. Also, equine babesiosis or piroplasmosis (*Babesia equi*, *B. caballi*) can be transmitted by this tick.

Hosts: Primarily equines (e.g., horses, mules, asses)

Description: Unlike most other *Dermacentor* species, *D. nitens* adults are inornate and dark brown to black in color. They have short mouthparts.

Biology/Behavior: *Dermacentor nitens* is a one-host tick. The parasitic period in which larvae, nymphs, and adults feed on the same host lasts from 4 to 6 weeks. All stages may be recovered on host animals year round in endemic areas.

Pacific Coast Tick (Order Acari; Family Ixodidae) (Figure 8.11)

Species Name: *Dermacentor occidentalis* (Marx)

Geographic Distribution: Confined to the Pacific Coast area from Oregon to California.

Veterinary Importance: Adult Pacific Coast ticks, which feed on larger animals, will attach onto their host over all body regions with no apparent site preference. They have been implicated in the transmission of bovine anaplasmosis.

Hosts: Immatures feed on a variety of small mammals. Adults will feed on cattle, horses, sheep, dogs, and humans.

Description: *Dermacentor occidentalis* resembles *D. variabilis* and *D. andersoni*, having a reddish-brown body, distinctive ornate scutum, and short mouthparts.

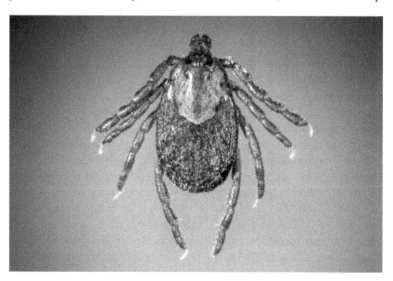

Figure 8.11
Pacific Coast tick female, Dermacentor occidentalis. *(Photo reproduced with permission by Jim Kalisch, University of Nebraska.)*

Biology/Behavior: This is a three-host tick with the immature stages feeding primarily on a variety of small mammals. Immatures are common in the spring and summer. Adults may be found throughout the year but reach peak populations in April and May within their distribution range.

Black-Legged Tick/Deer Tick (Order Acari; Family Ixodidae) (Figure 8.12)

Species Name: *Ixodes scapularis* (Say)

Geographic Distribution: Widespread in the eastern, south central, and midwestern United States and south through central Texas into Mexico.

Veterinary Importance: In endemic areas, livestock (e.g., cattle) can sometimes become heavily infested with adult *I. scapularis* during the winter and early spring. Infestations can be found on the axillaries, dewlap, tailhead, escutcheon, and udder/scrotum. This tick has commonly been found on other domestic animals, including horses, dogs, and cats. In recent years, this tick has been observed to be expanding its range from the northeastern United States and upper Midwest. It is being found more commonly infesting dogs and cats. Increases in and movement of white-tailed deer populations have significantly contributed to the reproductive success and spread of this tick.

Ixodes scapularis is the primary vector of *Borrelia burgdorferi*, the spirochete bacterium causative agent of Lyme disease. This disease is not only of importance medically as a human disease, but several domestic animals have shown to be susceptible to clinical signs. Dairy and beef cattle, horses, dogs, and cats have shown varied clinical symptoms of Lyme disease. In these animals, symptoms may include joint pain and swelling, lameness, fever, and lethargy.

Hosts: Immature stages normally feed on a variety of small mammals, birds, and lizards, with larvae most frequently feeding on the white-footed mouse, *Peromyscus leucopus*. Nymphs feed primarily on a variety of small, ground-dwelling vertebrates but will feed on humans and pets. Adults feed primarily on white-tailed deer but will be found on other animals, including cattle, horses, sheep, dogs, cats, a variety of other mammals, and humans.

Description: Adults are small, about 3 mm long, dark brown to black in color, with no white markings dorsally. In females, the abdomen behind the scutum is orange to dark red in color.

Biology/Behavior: *Ixodes scapularis* is a three-host tick. Larvae are most abundant from July into September. Once they successfully feed, they molt into nymphs. They often overwinter as unfed nymphs. Nymphs are most abundant from late spring into early summer. Once they feed and drop from their host,

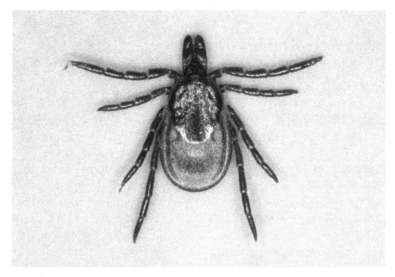

Figure 8.12
(Top) Black-legged tick female, Ixodes scapularis. *(Bottom) Black-legged tick male,* Ixodes scapularis. *(Photos reproduced with permission by Jim Kalisch, University of Nebraska.)*

they molt into adults. Adults are most prevalent in late summer into autumn, and in early spring if they overwinter unfed. Replete adult females drop from their host and lay about 3000 eggs before they die. The complete lifecycle usually takes 2 years depending on environmental conditions and host availability.

Brown Dog Tick (Order Acari; Family Ixodidae) (Figure 8.13)

Species Name: *Rhipicephalus sanguineus* (Lareille)

Geographic Distribution: This tick is considered to be the most widely distributed tick in the world, found throughout both tropical and temperate regions. It is primarily a pest of dogs and rarely feeds on other animals or humans. It is often found in kennels and inside residences in and around pet bedding. All feeding stages feed on dogs. Adults commonly attach on the ears and between the toes. Larvae and nymphs are often found along the back. In heavy infestations, ticks may be found on any part of the dog's body. Heavy infestations can result in skin irritation and damage.

In the United States, the brown dog tick is a known vector of *Ehrlichia canis* that causes canine ehrlichiosis, and *Babesia canis* that causes canine babesiosis in dogs. Symptoms of canine ehrlichiosis include lameness and fever. Canine babesiosis can cause fever, anemia, jaundice, and anorexia in dogs.

Hosts: Primarily dogs but occasionally other mammals and rarely humans

Description: Unfed adults are about 4 mm long and red-brown in color. They lack ornamentation. They have a distinctive hexagonal basis capituli.

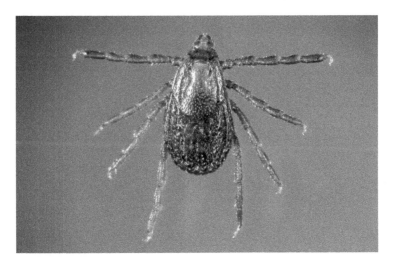

Figure 8.13
Brown dog tick, Rhipicephalis sanguineus. *(Photo reproduced with permission by Jim Kalisch, University of Nebraska.)*

Biology/Behavior: The brown dog tick is a three-host tick. Unlike other ticks, it can complete its entire lifecycle indoors. For this reason, it has been able to establish populations in more temperate and colder regions. Adult females feed for about 1 week. When replete, they drop off the host and find a secluded place to lay their eggs. Crack and crevice areas in residence, dog runs, and kennels are typical locations where these fed ticks can be found. A fully engorged female will begin laying eggs within 4 days and continue for up to 15 days. She will lay up to 5000 eggs before she dies. Larvae hatch in about 20 to 30 days and seek a host. After feeding for 3 to 7 days, they drop off their host and molt into nymphs in about 2 weeks. Once nymphs attach on a host, they feed for 5 to 10 days and then drop off to molt into adults. Adults emerge in about 2 weeks. The complete lifecycle can be completed in as short as 60 days, but it often takes longer. The brown dog tick is quite hardy and can survive from 3 to 5 months or longer within each stage without feeding.

ARGASID TICKS OF VETERINARY IMPORTANCE

Fowl Tick (Order Acari; Family Argasidae) (Figure 8.14)

Species Names:

Argas persicus (Oken)
A. radiatus (Raillet)
A. sanchezi (Duges)
A. miniatus (Koch)

Geographic Distribution: *Argas persicus* is widely distributed throughout tropical and warmer regions of the world but is primarily found in Old World

Figure 8.14
Fowl tick, Argas *sp. (Photo courtesy of Centers for Disease Control.)*

countries and is relatively rare in the United States. *Argas radiatus* is more common in the New World with *A. sanchezi* and *A. miniatus* sometimes encountered primarily in the western United States, Mexico, and Central and South America.

Veterinary Importance: Fowl ticks are not commonly found in modern, commercial poultry operations. Breeder houses, roost, and floor-raised poultry operations sometimes experience infestations. They are more common in warm climates and are found in greatest abundance during warm, drier seasons of the year. When present and when conditions are suitable, they can build up in large numbers. They feed primarily beneath the wings and sparsely feathered parts of the body. Infested birds experience weakness, slow growth, and decreased productivity of eggs and meat. Ruffled feathers, poor appetite, and diarrhea are typical symptoms of infestation. In heavy infestations, death may occur from exsanguination. Younger birds usually have a higher mortality. Tick paralysis in birds has also been reported from infestations of fowl ticks.

Fowl ticks are also capable of transmitting several pathogens to poultry. Included among these are the pathogenic spirochete *Borrelia anserine* and the blood protozoan *Aegyptianella pullorum*.

Hosts: Domestic fowl and several wild birds, including ducks, geese, and pigeons

Description: The fowl tick varies from yellowish-brown to reddish-dark brown in color depending on its stage of engorgement. Adult females are about 8 mm long by 5 mm wide, and males are about 6.5 mm long by 4.5 mm wide. Engorged females will expand to up to 10 mm long by 6 mm wide. The margin of the body is composed of irregular quadrangular plates (or cells).

Biology/Behavior: Fowl ticks are nocturnal with nymphs and adults being very active at night in seeking hosts. During the day, they hide in cracks and crevices of bird-nesting or roost areas. Adult females will lay a total of about 500 to 800 eggs in batches of 25 to 100 eggs, depositing them in crack and crevice areas in the poultry facility. They require a blood meal before laying each batch. Eggs hatch in 6 to 10 days during warm weather but may take up to 3 months in cooler weather. Larvae normally seek a host within 4 to 5 days and are active both day and night. They feed for 4 to 5 days and then drop from the host to molt into first-stage nymphs after about 1 week. These nymphs nocturnally seek a host and feed only for about 10 to 45 minutes. They then leave the host again and molt into second-stage nymphs several days later. Within 1 to 2 weeks, these second-stage nymphs find a host and feed for about 1 hour and leave. There can be an additional one to two more nymphal stages, each taking separate short blood meals. Emerged adults seek a host about 1 week later, and females begin to oviposit 3 to 5 days after mating. The complete lifecycle takes from 4 to 8 weeks under favorable warm conditions but can be much longer in colder weather or if there is a lack of suitable hosts. There are normally from one to ten generations per year. Larvae and nymphs can survive several months without feeding, and adults can survive for 1 or more years in the absence of hosts to feed upon.

Pajaroello Tick (Order Acari; Family Argasidae) (Figure 8.15)

Species Name: *Ornithodoros coriaceus* (Koch)

Geographic Distribution: Pacific Coast from Southern Oregon, through California, into Mexico.

Veterinary Importance: Primarily a pest of cattle and deer, but will feed on humans. It is most prevalently associated with animal bedding areas, especially those of deer and cattle on rangeland. Tick activity is stimulated by host presence. These ticks will attach onto a host at all hours of a day. In humans, the attachment "bite" of this tick is quite painful and is probably quite irritating to host animals. A discoloration and inflammation often develop around the attachment site. Of most concern with cattle, *O. coriacus* has been incriminated as a primary vector of the spirochete that causes epizootic bovine abortion (EBA). This disease is most prevalent in the foothills of California, with some cases in cattle also reported in Nevada. This tick has also been associated with cases of tick paralysis.

Figure 8.15
Pajaroello tick, Ornithodoros coriaceus. *(Photo reproduced with permission by Carmen Guzman-Cornejo, University of California.)*

Hosts: Primarily cattle and deer but occasionally feeds on humans and other animals

Description: The pajaroello tick is about 1 cm long with an irregular oval shape, thick margin, and roughly shagreened. Its color is somewhat earthy-tan to a spotted rusty red.

Biology/Behavior: The pajaroello tick is a multihost tick. Larvae often stay attached on a host for 10 to 20 days. They molt into first-stage nymphs that do not feed. The second-stage nymphs, and subsequent nymphal stages (from three to seven), feed on a separate host for a short period of 15 to 30 minutes. Adult ticks will feed on several different hosts for 15- to 30-minute periods. Adult females deposit small batches of eggs after each feeding. Depending on environmental conditions and host availability, a complete lifecycle from egg to adult takes from 1 to 2 years. Adults have been known to survive for 2 to 5 years.

Spinose Ear Tick (Order Acari; Family Argasidae) (Figure 8.16)

Species Name: *Otobius megnini* (Duges)

Geographic Distribution: From western Canada, western and southwestern United States, Mexico, Central and South America, Africa, and India

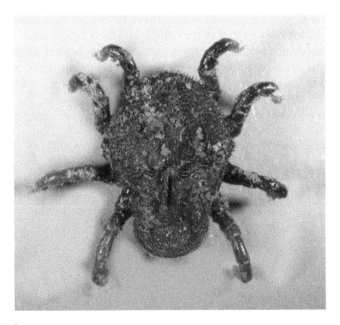

Figure 8.16
Spinose ear tick, Otobius megnini. *(Photo reproduced with permission by Michael Dryden, Kansas State University.)*

Veterinary Importance: The spinose ear tick can be of concern, especially in cattle and horses. Larvae and nymphs of this tick feed in the external ear canal of their host, remaining attached for several months. They cause pain and irritation, animals develop flop-ears, and there is a noticeable foul-smelling discharge of wax and tick debris. Infested animals become restless and are often observed shaking their heads. This tick is not known to transmit disease, but infestations have predisposed animals to attack by screwworms and other blow flies.

Hosts: Cattle, horses, other equine, sheep, cats and dogs, wild canines, rabbits, elk, mountain sheep, and goats

Description: The parasitic nymphs are readily identified by the presence of spines on their integument. In adults, the integument is granulated. Adults range in size from 4 to 8 mm long. The hypostome is much reduced in the nonparasitic adults but is well developed in nymphs.

Biology/Behavior: The spinose ear tick is a one-host tick. Newly emerged larvae crawl up on vegetation, contact a suitable host, and crawl up on the host animal and into the inner folds of the outer ear where they attach and feed. They molt in the ear to the first-, and then to second-stage nymphs. The second-stage nymphs take on the distinctive spiny appearance. From 5 weeks to several months may be spent on the host during the parasitic larval and nymphal stages. During this time, these ticks may be transported great distances by their host animal. After completing their blood meal, the second-stage nymphs detach from their host and drop to the ground where they molt again and then molt into adults. Adults do not feed. Adults mate within a day or two, and oviposition occurs in 2 to 4 weeks. Adult females will lay about 500 eggs, and larvae emerge in about 18 to 24 days. Although these ticks can be found on hosts all year, they are most prevalent in late winter and spring.

————————————CHAPTER 9

OTHER ARTHROPOD GROUPS OF VETERINARY IMPORTANCE

TABLE OF CONTENTS

There is an array of other arthropods that can be of veterinary importance associated with livestock production and/or the health and well-being of companion animals. Some species have a direct impact on animals (e.g., fire ants, bed bugs, blister beetles), some may be detrimental to production practices (e.g., wood-destroying beetles), and some are considered beneficial and are often used as viable components of integrated pest management programs (e.g., dung beetles, predatory beetles, certain species of parasitic wasps). Included here are selected insect groups that may be encountered in conjunction with livestock and companion animals.

COCKROACHES (ORDER BLATTARIA)

Cockroaches are among the oldest insect groups. There are about 4000 species worldwide. A few species of cockroaches have adapted to living in human and animal habitations. Although they are primarily nuisance pests, their presence in and around livestock production and processing facilities and companion animal quarters can have important health implications. Cockroaches feed on a wide variety of organic material, including animal products and products produced from animal production and processing, causing potential contamination by pathogens associated with their feeding habits. Cockroaches have also been shown to serve as intermediate hosts of parasites and pathogens that can affect domestic animals.

Oriental Cockroach (Order Blattaria; Family Blattidae) (Figure 9.1)

Species Name: *Blatta orientalis* (Linnaeus)

Geographic Distribution: Europe, Asia, Africa, North and South America

Veterinary Importance: This cockroach is not considered of significant veterinary importance. It sometimes will be found in confined livestock and poultry facilities in wet/damp areas.

Description: Adults are dark brown to black in color and 25 to 30 mm long. Adults have reduced wings, shorter than the abdomen, and they cannot fly.

Biology/Behavior: Adult females produce about 14 to 15 egg capsules averaging 12 to 16 eggs each. Emerged nymphs molt from 7 to 10 times and take several months to a year to complete development. Adults may live for several months. Their movement is fairly restricted and confined to moist areas. They are most active at night and are rarely seen during daylight hours.

Figure 9.1
Oriental cockroach, Blatta orientalis. *(Photo courtesy of John Obermeyer, Purdue University.)*

American Cockroach (Order Blattaria; Family Blattellidae) (Figure 9.2)

Species Name: *Periplaneta americana* (Linnaeus)

Geographic Distribution: Probably worldwide

Veterinary Importance: The American cockroach is found in a wide variety of habitats and is often found infesting confined livestock and poultry production facilities and processing facilities as well as residences. It is capable of flying and is quite mobile. They feed on a variety of food sources but prefer decaying organic matter. This cockroach has the potential for the dissemination of several pathogenic microorganisms that can be of veterinary concern (e.g., *Salmonella* sp., *Escherichia coli*).

Description: The American cockroach is a large species averaging 35 to 50 mm long. It is reddish-brown in color with light and dark patterns on its pronotum. Adults are fully winged and capable of flying.

Biology/Behavior: Adult females produce about 6 to 14 egg capsules averaging 12 to 16 eggs each during their lifetime of up to 15 months to 2 years. They

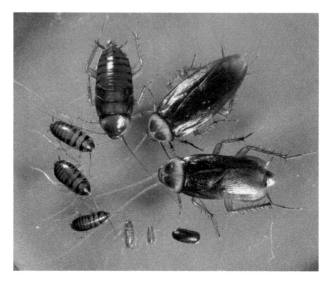

Figure 9.2
American cockroach, Periplaneta americana. *(Photo courtesy of John Obermeyer, Purdue University.)*

drop each egg capsule within a day after it is formed. Nymphs emerge in 50 to 55 days and will complete their development in 13 to 14 months, molting from 9 to 13 times.

German Cockroach (Order Blattaria; Family Blattellidae) (Figure 9.3)

Species Name: *Blattela germanica* (Linnaeus)

Geographic Distribution: Widely distributed worldwide

Veterinary Importance: The German cockroach is considered a major domestic pest species and the most common cockroach found in human habitation in most areas where it occurs. In the United States, it is considered the most economically important urban pest. It is not generally commonly found in livestock production facilities. However, it has been incriminated as a vector of certain pathogens of veterinary importance. It has been shown to be associated with the transmission of esophageal nematodes *Physaloptera rara* and *P. praeputialis* that can affect dogs, cats, and other animals. In poultry, German cockroaches have been incriminated as intermediate hosts of various stomach nematodes (e.g., *Tetrameres americana, T. fissispina,* and *Cyrnea colini*). Also, the gullet worm of cattle, *Gongylonema pulchrum,* has been shown to undergo development in the German cockroach, but its role in possible transmission is not known.

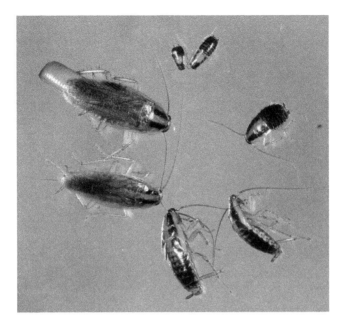

Figure 9.3
German cockroach, Blattela germanica. *(Photo courtesy of John Obermeyer, Purdue University.)*

Description: Adults are pale to medium brown in color, about 16 mm long, and have two dark, longitudinal stripes on the pronotum. They have well-developed wings but do not fly. Nymphs resemble adults except they are smaller, wingless, darker in color, and have a single light stripe running down the middle of their back.

Biology/Behavior: Adult females produce about four to eight egg capsules averaging 30 to 48 eggs each during their lifetime. These egg capsules are produced about every 20 to 25 days. The female carries the egg capsule with her until just a few hours before the nymphs emerge. Nymphs molt six to seven times and complete their development in 7 to 12 weeks. Adult females molted from nymphs are capable of egg development within 7 to 10 days.

Brownbanded Cockroach (Order Blattaria; Family Blattellidae) (Figure 9.4)

Species Name: *Supella longipalpa* (Fabricius)

Geographic Distribution: In tropical and subtropical areas of the world and in temperate regions of Europe and North America, primarily indoors in warmer climates.

Figure 9.4
Brownbanded cockroach, Supella longipalpa. *(Photo courtesy of John Obermeyer, Purdue University.)*

Veterinary Importance: This species is rarely associated as being a veterinary pest. They are more commonly found in homes, apartments, hotels, and hospitals. They may occasionally be transported into confined livestock or poultry operations with crates or supplies but usually do not get established.

Description: Adults are similar in size to German cockroaches, about 13 to 15 mm long. They can be distinguished by the adults having two light, transverse bands across the base of the wings and abdomen, and in the same position of the unwinged nymphs. Adults may fly if disturbed.

Biology/Behavior: Adult females produce about 14 egg capsules containing about 18 eggs each during their lifetime of up to 10 months. They attach their egg capsules onto furniture, bedding, or other dry substrates. Eggs hatch in 50 to 75 days, and nymphs mature in about 160 days. Brownbanded cockroaches are quite active, especially at night, and are often observed on walls and ceilings.

Surinam Cockroach (Order Blattaria; Family Blaberidae) (Figure 9.5)

Species Name: *Pycnoscelus surinamenisis* (Linnaeus)

Distribution: Found in tropic areas around the world, occurs in the southern United States from North Carolina southward.

198

Figure 9.5
Surinam cockroach, Pycnoscelus surinamenisis.

Veterinary Importance: This cockroach is sometimes encountered in compost piles. It has been shown to be the intermediate host for poultry eye
worms *Oxyspirura mansoni* and *O. parvorum.* These parasites can cause
symptoms from mild conjunctivitis to severe vision impairment.

Description: This cockroach is about 18 to 25 mm long with shiny dark
brown wings and a black body. The front edge of the pronotum has a pale
white band.

Biology/Behavior: This cockroach has a unique lifecycle. The egg capsule
is retained inside the abdomen from which nymphs emerge. Also, it repro-
duces parthenogenetically in North America, producing only female offspring.
Elsewhere, both males and females are produced. Females live for about 10
months and produce around three egg capsules, each containing an average
of 24 eggs. Adults are active at night and will often be seen flying around
light sources.

TRUE BUGS (ORDER HEMIPTERA)

Of the true bugs, all of which possess piercing–sucking mouthparts, a few species are of veterinary importance as being hematophagous, feeding on blood.
Of primary importance are the triatomine species in the family Reduviidae
which are capable of transmitting the pathogen causing trypanosomiasis and
bugs in the family Cimicidae (e.g., bed bugs, bat bugs) which can become
established in poultry operations.

Figure 9.6
Triatomine bug.

Triatomine Bugs (Order Hemiptera; Family Reduviidae) (Figure 9.6)

Species of Veterinary Importance in the Americas:

Triatoma infestans (Klug)
Rhodnius prolixus (Stal)
Panstongylus geniculatus (Latreille)
Panstongylus megistus (Burmeister)
Others

Veterinary Importance: Triatomine bugs transmit the causative agent of Chagas disease, *Trypanosoma cruzi*, to a variety of domestic and wild animals (e.g., canines, opossums, armadillos, raccoons, rodents, and others). Canine trypanosomiasis has been reported in central and South America and in parts of Texas. Triatomine infestations in poultry operations in Central America have been reported to cause blood loss and affect productivity in chickens.

Hosts: Wide variety of domestic and wild animals

Description: Triatomines average around 25 mm long. They are dark brown to black in color with patterns of yellow to orange or red markings on abdominal margins. They have narrow heads with their piercing–sucking mouthparts folded under their head when not feeding.

Biology/Behavior: All life stages feed on blood. Adult females mate within 1 to 3 days upon molting from mature nymphs. After mating, females begin oviposition in 10 to 30 days, depositing one to two eggs daily (10 to 30 eggs

200

between blood meals). Females will produce from 100 to 1000 eggs in their lifetime, depending on species. Eggs are deposited on various substrates, often in and around habitats close to suitable hosts. Emerged nymphs molt several times. The number of molts varies by species, environmental conditions, and host availability. The entire lifecycle takes from 3 to 4 months to 1 year or longer.

Bed Bugs, Bat Bugs (Order Hemiptera; Family Cimicidae) (Figure 9.7)

Species of Veterinary Importance

Cimex lectularius (Linnaeus)—Bed bug
Cimex pilosellus (Horvath)—Western bat bug
Cimex adjunctus (Barber)—Eastern bat bug

Geographic Distribution: *Cimex lectularius* is cosmopolitan, *C. pilosellus* is found in western North America, and *C. adjunctus* is found from Colorado to the Atlantic Coast.

Veterinary Importance: Bed bugs and bat bugs can be major pests in commercial poultry operations. They will feed on poultry at night and hide in crack and crevice areas and debris in poultry houses during the day while digesting

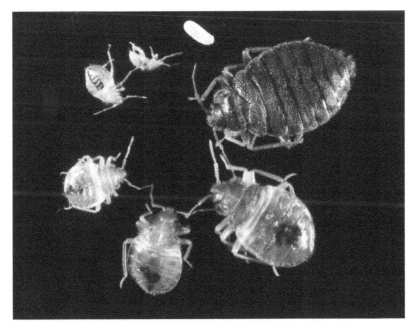

Figure 9.7
Bed bug, Cimex lectularius. *(Photo courtesy of John Obermeyer, Purdue University.)*

their blood meals. They are more frequently encountered in floor-raised poultry and roost facilities rather than in caged systems. Caged-layer operations lack adequate hiding places and bird access by these bugs. In heavily infested facilities, birds will show weight loss, lowered egg production, and decreased feed efficiency. Young birds may be killed from excessive blood loss. Poultry farm workers may be bitten by both bed bugs and bat bugs. They may also be carried by workers into their private residences to establish infestations there. These bugs are not known to be of any significance in disease transmission to humans or poultry.

Hosts: Humans, domestic poultry, bats, other birds, swine, other domestic animals, and rodents

Description: *Cimex* species are dorsoventrally flattened and oval in appearance. Adults range from about 5 to 7 mm long. They have very short, stubby, nonfunctional wing pads. Unfed, they are brown in color, and they turn red and plump after taking a blood meal. They have short, beak-like piercing–sucking mouthparts. Nymphs resemble adults but are smaller and of a pale yellow color. Bed bugs and bat bugs look very much alike. A distinguishing characteristic to tell them apart is that in bat bugs the fringe hairs on the pronotum are as long as or longer than the width of the eye. In bed bugs, these fringe hairs are shorter.

Biology/Behavior: Bed bugs and bat bugs have similar lifecycles. Development from egg to adult ranges from 2 to 15 weeks or longer, depending on environmental conditions and host availability. Multiple generations usually occur per year. Mated females feed to repletion on blood and begin to lay eggs 3 to 6 days later. They feed about every 3 to 4 days, laying 6 to 10 eggs (2+ per day) between blood meals. From 200 to 500+ eggs will be laid in their lifetime. Eggs are usually laid in crack and crevice hiding places where the bugs rest during the day. Eggs hatch in 6 to 12 days. There are five nymphal stages, each stage lasting from 2 to 10 days. Adults can live for 10 to 12 months or longer under favorable environmental conditions. These bugs have a distinct odor given off from thoracic glands. Also, in heavier infestations, their daytime hiding places are often cluttered with nymphs, adults, fecal material, cast "skins," and eggs. They leave these hiding places primarily to seek a host to feed upon. Warmth and carbon dioxide are key factors they migrate to in finding a host.

BEETLES (ORDER COLEOPTERA)

Beetles represent the largest order of insects. The vast majority of beetles have little or no veterinary importance. There are a few groups, however, that do have an impact on domestic animals. Some beetles are detrimental in one way or another, and others are beneficial. Of those that are detrimental, some harm animals directly, such as being toxic when ingested (e.g., blister beetles). Others may serve in the transmission of disease agents. Also, some beetles are of economic concern because they cause structural damage to confined livestock and poultry facilities (e.g., dermestid beetles).

Beetles can also be beneficial. Some species play an important role in altering animal waste, making it unsuitable for pest species development. Such is the case with dung beetles destroying pasture cattle manure making it unavailable for pasture fly development. Other beetles are beneficial in that they are predators of pest insects and can be utilized as part of a biological control program. An example of this is the histerid beetle, *Carcinops pumilio*, that is an effective predator of house flies.

Detrimental Beetles

Blister Beetles (Family Meloidae) (Figure 9.8)

Species of Veterinary Importance:

Epicauter sp., including
   E. vittata (Fabricius)—Striped blister beetle
   E. pennsylvanica (De Geer)—Black blister beetle
   E. pestifera (Werner)—Margined blister beetle

Figure 9.8
Striped blister beetle, Epicauter vittata.

E. maculata (Say)—Spotted blister beetle
Others

Geographic Distribution: Blister beetles occur worldwide. Of pertinence as veterinary pests in the United States, *E. vittata* occurs in the eastern states, *E. pennsylvanica* is found across most of the country, *E. pestifera* is most commonly found in the south-central states, and *E. maculata* occurs in the western states.

Veterinary Importance: Blister beetles can be of concern when beetle-infested hay is fed to livestock and horses. The blister beetle species listed above are frequently found feeding in alfalfa. Animals that ingest blister beetle–infested alfalfa hay are prone to being affected. Blister beetles contain the toxic compound cantharidin that causes blistering to skin tissue. When ingested, it can cause various symptoms, including inflammation, colic, fever, depression, increased heart rate and respiration, dehydration, sweating, and diarrhea. Extremely affected animals can die. Horses are especially susceptible. Cantharidin is very stable and remains toxic even in dead beetles. Levels of cantharidin vary by species. *Epicauter vittata*, as an example, has about five times more cantharidin than *E. pennsylvanica*. Only 1 mg of cantharidin per kg of horse weight is enough to kill a horse. Individual beetles, depending on species, contain from about 0.1 to 11 mg of cantharidin. As few as 25 beetles (dead or alive) of some species to over 1000 beetles of other species are enough to kill a horse. Even as few as five beetles ingested can cause colic. Horses are most frequently affected because they are often fed the type of forage most likely infested. Ruminant animals such as cattle, sheep, and goats fed infested hay can also be affected.

Hosts: Blister beetles feed on a variety of plants. When found in alfalfa used for hay, animals that are fed this can be affected, including horses, cattle, sheep, and goats.

Description: Blister beetles have a long slender body (2 to 3 cm long), narrow pronotum, broad head, and long antennae about one-third the length of their bodies. Their front wings are soft and flexible, as compared to the harder front wings of most beetles. The coloration of blister beetles varies by species. Of the more common species, the striped blister beetle has orange and black stripes on the front wings; the black blister beetle is solid black in color; the margined blister beetle is black with a narrow gray stripe around the perimeter of the front wings; and the spotted blister beetle has black spots on the front wings on a background of dense, short gray hairs.

Biology/Behavior: Adult blister beetles are active in the summer, being more active from mid-July to early August. The highest populations are usually seen from third or fourth cuttings of alfalfa. Adults feed on leaves at the top of plants and are attracted to flowers to feed on nectar and pollen. Females lay their eggs, often in clusters, in the soil near their food source in late summer. Larvae that emerge from these eggs are quite active and crawl over the soil surface seeking their preferred food source of grasshopper eggs. Once they find an adequate grasshopper egg mass, they settle at this site to complete their development, becoming grub-like in appearance with rudimentary legs. They overwinter as mature larvae and pupate in the spring and summer. Adults emerge soon after. Because of their dependency on grasshoppers, blister beetle populations will fluctuate with population levels of grasshoppers. Blister beetle numbers often increase dramatically following a year of high grasshopper populations.

Lesser Mealworm (Order Coleoptera; Family Tenebrionidae) (Figure 9.9)

Species Name: *Alphitobius diaperinus* (Panzer)

Geographic Distribution: Cosmopolitan

Veterinary Importance: *Alphitobius diaperinus*, often referred to as darkling beetles or litter beetles, are often found in abundance in the litter of poultry houses, such as in floor-raised turkey and broiler operations, and in the accumulated wastes in caged egglayer operations. On one hand, these beetles can be considered beneficial in caged egglayer operations, in that they compete in the same habitat as house flies and can help aerate and dry manure, making it unsuitable for house fly development. On the other hand, they can cause extensive damage as mature larvae bore into structural materials seeking areas

Figure 9.9
Lesser mealworm adult/larva, Alphitobius diaperinus. *(Photo reproduced with permission by Univar.)*

to pupate and complete their development, especially into walls and insulation material. Damage in poultry houses can cause increased heating/cooling costs and additional building repair costs when the damaged areas are repaired. Energy costs in beetle-damaged poultry houses have been reported to be 67% higher than in undamaged houses. Also, birds will consume more feed when environmentally stressed. They are also known as potential vectors of several poultry disease pathogens (e.g., viruses that cause acute leukosis or Marek's disease, Gumboro disease, turkey coronavirus, Newcastle disease, avian influenza; bacteria such as *Salmonella typhimurium*, *Escherichia coli*, *Aspergillus* spp., and *Staphylococcus* spp.; protozoans such as *Eimeria* spp. that cause coccidiosis; pathogenic fungi; helminthes such *Subulura brumpti*; and fowl cestodes). Large beetle populations may also become a public nuisance at cleanout time because of adult beetle migration from fields where poultry waste is spread into nearby residential properties.

Hosts: Often found in abundance in poultry operations, in the litter of turkey and broiler operations, and in accumulated manure under caged egglayer operations.

Description: *Alphitobius diaperinus* adults are dark brownish to black in color and about 0.6 cm long. The elytra, or front wings, have moderately grooved striations. The underside is dark reddish-brown. Larvae are wireworm-like with a segmented body and abdominal tip tapering posteriorly, and they have three pairs of legs. Mature larvae are about 7 to 11 mm long.

Biology/Behavior: Adult female beetles, after mating, will lay from 200 to 400 eggs in cracks and crevices in poultry facilities or in the manure or litter. Adults can live from 3 to 13 months, with females depositing eggs at 1- to 5-day intervals throughout their lives. Larvae hatch in 4 to 7 days and complete their development in 40 to 100 days. Larvae spend most of their time in the manure or litter feeding on damp and moldy grain sources. They will also feed on nutrients in manure and litter, dead or sick birds, and cracked and broken eggs. There are from 6 to 11 larval instars. Upon maturity, larvae migrate to find secluded pupation sites. They often tunnel into walls and chew holes in Styrofoam, fiberglass, and polystyrene insulation panels to pupate. When adults emerge, they will do further damage when they leave these pupation sites. Larvae and adults are primarily nocturnal and most active at dusk. When disturbed, they quickly burrow into the litter. They are active year round but developmental time is long during the colder months.

Dermestid Beetles (Order Coleoptera; Family Dermestidae)

Species of Veterinary Importance

Dermestes maculates (De Geer)—Hide beetle (Figure 9.10)
Dermestes lardarius (Linnaeus)—Larder beetle (Figure 9.11)

Figure 9.10
Hide beetle adult/larva, Dermestes maculatus. *(Photo reproduced with permission by Univar.)*

Figure 9.11
Larder beetle adult, Dermestes lardarius. *(Photo reproduced with permission by Joseph Berger, Bugwood.org.)*

Geographic Distribution: Worldwide

Veterinary Importance: These dermestid beetles are of similar veterinary importance as *Alphitobius diaperinus* when they occur in poultry production facilities. Their feeding habits are different in that they feed on bird carcasses, skins, hides, feathers, dead insects, and other animal and plant products. Broken eggs and dead bird accumulations in manure enhance hide beetle populations. Also, mature larvae often bore into wood posts, beams, and paneling to pupate and, over time, can cause severe structural damage to a building. There have been instances of building collapses due to damage caused by dermestid larval burrowing activity. Even partial stress to a building can cause equipment failure from cage and feed system warping and shifting.

Hosts: See *Alphitobius diaperinus*. They can also sometimes be found around animal and poultry processing facilities.

Description: These dermestid beetles are larger than *Alphibobius diaperinus* (0.8 cm long). The larder beetle is dark brown with the basal halves of the front wings densely covered with coarse, pale yellow hairs. There are six dark spots within the band of yellow. The underside of the body and legs are covered with finer yellow hairs. The hide beetle is similar in shape to the larder beetle, but the front wings are covered in black hairs with no other coloration. The underside of the body is mostly white. The larvae of both species are longer than the adult beetles (up to 1.3 cm long), reddish brown to black in color, slender, densely covered with short and long hairs, and with two spines on top near the posterior end. In larder beetle larvae, these spines curve backward. In hide beetle larvae, these spines curve forward.

Biology/Behavior: Mated adult females deposit eggs in batches of six to eight at a time on food sources or in sheltered places nearby. From 200 to 800 total eggs may be deposited by an individual female. Emerged larvae feed and develop to maturity in 40 to 60 days. Mature larvae migrate in search of a pupation site. They will bore into various hard surfaces to pupate, usually preferring softwood. When they bore into posts, studs, and rafters, they can "honeycomb" these structures and cause structural weakening. The pupal stage lasts for about 2 weeks. Emerged adults soon mate and egg laying starts. The complete lifecycle is completed in 60 to 80 days. Indoors, these beetles may breed continuously throughout the year.

Beneficial Beetles

Histerid Beetles (Order Coleoptera; Family Histeridae) (Figure 9.12)

Species of Veterinary Importance: *Carcinops pumlio* (Erichson)

Figure 9.12
Histerid beetle, Carcinops pumilio.

Veterinary Importance: *Carcinops pumilio* is an effective predator of the house fly, *Musca domestica,* in confined livestock facilities, especially in caged egglayer operations with in-house storage of manure. Both adults and larvae are predators. Adult *C. pumilio* beetles have been reported to consume as many as 24 house fly eggs per day and up to 24 fly larvae (mostly first instar) per day. They often make up an important component of a fly pest management program and have also been made available for use commercially. Of concern, however, as with many other insects occupying animal wastes, these beetles have been found to be associated with the transmission of *Salmonella* sp. and such parasites as tapeworms.

Hosts: Found in accumulated animal and poultry wastes where there are ample insect food sources.

Description: Adult beetles are small (2 to 3 mm long), broadly oval, and shiny dark brown to black. Their front wings (elytra) cover all but the end of the abdomen. Also, their antennae have orange-colored clubs at the ends.

Biology/Behavior: Adult females lay their eggs in relatively dry manure. Larvae develop through two to three stages, and then pupate. The entire lifecycle takes about 1 month. Adults can live for several weeks. In a caged egglayer house, it may take several weeks for significant beetle populations to develop. Fresh poultry manure is very moist, and these beetles don't usually build up in numbers until after some drying occurs in the accumulating manure pile.

Dung Beetles (Order Coleoptera; Family Scarabaeidae) (Figure 9.13): Manure from pastured raised cattle is the primary breeding source for horn flies, *Haematobia irritans,* and face flies, *Musca autumnalis,* two flies of significant economic importance to the cattle industry in North America. Cattle manure in pastures and rangeland also attracts other insects for its nutrient value. One

Figure 9.13
Dung beetle, Onthophagus gazella. *(Photo courtesy of Patrick Jones, Purdue University.)*

group of insects includes dung beetles in the family Scarabaeidae. Several species of dung beetles are commonly found in North American pastures. Efforts have been made to import Australian and African species of dung beetles to effectively complement native dung beetle populations in aiding in fly control. One species, *Onthophagus gazella*, the Afro-Asian dung beetle, has been successfully established in the southern states from South Carolina to California. Another species, *Euroniticellus intermedius*, imported from Australia, has gotten established in regions of Texas. This species is active in dry weather when native beetles are not and has been quite effective in destroying dung pats, making them unsuitable for horn fly development. Other species in the genera *Onthophagus* and *Apodius* include beetles that are commonly observed.

Dung beetles will destroy dung pats in different ways. Some species consume the dung pat and burrow into the soil under the pat. Some species occupy the dung pat, consuming the manure and depositing their eggs in the pat. Others are dung rolling beetles and break the dung pat into broad balls and roll them to suitable sites to bury them to nourish their young. In addition to making the manure unsuitable for fly development, dung beetle activity increases pasture yields resulting from incorporation of organic matter into the soil; reduces the infective stages of gastrointestinal parasites of livestock due to dung pat destruction; and causes greater pasture water retention and less runoff and improved plant root penetration and soil aeration due to beetle tunneling activity in the soil.

BEES, WASPS, AND ANTS (ORDER HYMENOPTERA)

In this large order of insects, bees, wasps, and ants can be found. Of concern to human and animal health are aculeate, or stinging Hymenoptera. In

Figure 9.14
Imported fire ant, Solenopsis *sp. (Photo reproduced with permission by Pest and Diseases Image Library, Bugwood.org.)*

these Hymenoptera, the females are equipped with a specialized oviposi-tor adapted for stinging which can pierce the skin and inject a venom. This stinger is used to subdue prey or can be used as a defensive weapon. Social Hymenoptera species that live in colonies often inflict the more serious or painful stings.

Most Hymenoptera have little veterinary importance. There have been reports of bees and wasps stinging domestic animals. Dogs on occasion have had fatal encounters from allergic reactions to honey bees and yellowjackets. In cattle, yellowjackets have been reported to injure teats of milking cows and the flesh of horses by biting in their foraging activity. In the United States, however, the biggest veterinary concern is with imported fire ants, *Solenopsis invicta* and *S. richteri* (Figure 9.14). These ants, common in the southern United States, are frequently found in pastures. Colonies of these ants can have more than 10,000 workers, with several colonies often found per acre in certain areas (Figure 9.15). When a colony is disturbed, ants will swarm over the intruder resulting in numerous stings. The venom of the fire ant produces an acute "burning" sensation. Newborn calves in infested pastures have some-times been killed by fire ant attack.

Some Hymenoptera have shown to be beneficial and have been used as biological control agents for house flies, *Musca domestica*, and stable flies, *Stomoxys calcitrans*. In the family Pteromalidae, there are several species that have shown to be of value in the biological control of these flies. These para-sitic wasps, called parasitoids, target fly pupae in which the female wasp drills a hole and deposits one or more eggs (Figure 2.5). When the eggs hatch, the larval wasps consume the developing fly in the puparium. When they com-plete development, the wasp chews its way out of the dead host's puparium and flies away as an adult. Common species of pupal parasitoids are in the genera *Muscidifurax* and *Spalangia*. Several commercial insectaries offer these parasitoids for sale for use in fly pest management programs. In the family

Figure 9.15
Fire ant mound. (Photo courtesy of U.S. Department of Agriculture, Animal and Plant Health Inspection Service [USDA APHIS] PPQ Archive, USDA APHIS PPQ, Bugwood. org.)

Encyrtidae, the parasitoid *Tachinaephagus zealandicus* deposits eggs in fly larvae and then emerge from the fly after they have pupated. This species, from South America, has been getting established in the United States.

CHAPTER 10

MANAGEMENT OF CATTLE PESTS (RANGE, CONFINEMENT, DAIRY)

TABLE OF CONTENTS

According to the U.S. Department of Agriculture (USDA), there are over 800,000 beef producers in the United States who are responsible for more than 100 million head of beef cattle. Over 35 million beef cows spend most of their lives on pastures or range land. At any given time, there are 14 million head in feedlots, and over 34 million head are slaughtered annually for meat production. Annual revenues in the beef industry exceed $38 billion. There are over 75,000 dairy enterprises in the United States with over 9 million milk cows. Annual revenues from dairy production exceed $36 billion.

Arthropod pests cause North American beef and dairy producers millions of dollars annually due to performance loss in reduced weight gains and milk production, reduced feed efficiency, veterinary costs for treatment of arthropod-borne diseases, direct physical harm or damage caused by arthropod attack, and the costs of control in reducing pest population levels. In this chapter, discussion will focus on pest management strategies in controlling the various arthropods that affect cattle production.

FLY PESTS

Pasture Flies

Of the flies occurring on pastured cattle, horn flies, *Haematobia irritans*, are of most concern, being found in all of North America (Figure 10.1). Face flies, *Musca autumnalis*, are also of major importance (Figure 10.2). Face flies are found coast to coast in temperate regions of the country. Control of these two pasture flies follows similar techniques.

Surveillance/Diagnosis: In monitoring horn fly and face fly infestations, direct visual fly counts on cattle are most practical. Fly activity on cattle will fluctuate throughout a season and during the course of a day. A rule of thumb is to make counts at the same time of the day, preferably between 10:00 A.M. and 2:00 P.M. Adult horn flies are obligate external parasites that stay on cattle continuously, except when females briefly leave to deposit eggs in fresh manure. They generally can be observed resting or feeding from the shoulder and sides of the animal to the tailhead. During hotter, sunny days or during rainy weather, they tend to congregate on the underside on the belly of cattle. Face flies, on the other hand, will be observed on the face and head of cattle. However, they spend only about 10% of their time actually on the animal. The rest of the time they are resting in surrounding vegetation or on fences. When

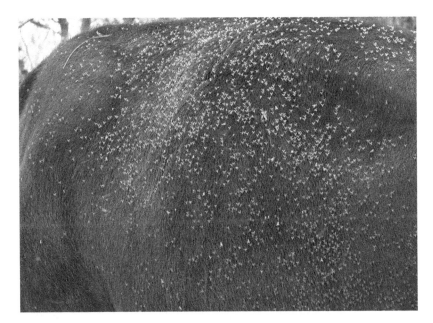

Figure 10.1
Horn flies on cow.

Figure 10.2
Face flies on cow.

face fly populations are high, they can also be found on the shoulders, backs, and sides of cattle.

Fly counts can be made with or without binoculars, depending on how close one can get to the animals. Horn flies should be counted on one or both sides of the animal. If flies are congregated underneath, a gentle disturbance to get the animal to move will usually cause flies underneath to "flush" and be counted. An estimate of the number of horn flies on the animal should be made in a set time period, preferably in a 15- to 30-second time interval. Face flies can be counted by counting the face flies landed on the face and head, also during a 15- to 30-second time interval. A representative sample of cattle in a herd should be used for counts, generally 10 or more individuals.

Control efforts are generally recommended when horn fly numbers average 200 or more per animal. This is considered by many investigators to be the "economic injury level." For face flies, if animals are visibly annoyed, or if there are signs of pinkeye incidence, this can be used to help in making a decision for control. Most often, however, where face fly populations are present, horn flies are also on cattle, and the decision to initiate control procedures is often dictated by horn fly infestation levels.

Pasture Fly Control

Cultural/Mechanical Control: Because horn flies and face flies develop in individual dung pats in pastures, cultural manure management practices are not practical. As a mechanical control of horn flies, a walk-through fly trap, first designed in the 1930s, has shown promise as an alternative to chemical horn fly

Figure 10.3
Walk-through horn fly trap. (Photo reproduced with permission by Robert Hall, University of Missouri.)

control (Figure 10.3). This trap is used where cattle must walk through it daily to obtain water, salt/mineral, or go from one pasture to another. On each side of the trap inside are inverted cone slats where the flies go in and cannot escape. As cattle enter the trap, strips of canvas hanging down from the inside top brush along the backs of the animals to disturb and cause the flies to leave the animal. The flies are attracted to light between the slats, enter the opening between the slats, and get caught in a screened enclosure. Recent field trials with this trap have shown effective horn fly control of up to 75% reduction. It has not, however, shown to effectively reduce face fly numbers. Information on this trap and how to construct it is available from the University of Missouri Extension Publication G1195: "Walk-Through Trap to Control Horn Flies on Cattle" by Robert D. Hall, Department of Entomology. The horn fly trap blueprint, University of Missouri Extension Publication MX1904C6, is also available.

Biological Control: Biological control of horn flies and face flies is limited currently to organisms that occur naturally in the pasture environment. Biological control agents that have been used successfully for confinement pest flies (e.g., stable flies, house flies) are not effective in pasture situations with isolated, individual dung pats. Dung beetles in the family Scarabaeidae can be of benefit by aiding in the destruction of the dung pat, making it unsuitable for fly development. Native dung beetle populations have not been effective in reducing fly populations, however. Efforts by the USDA in introducing Australian and African dung beetle species do show promise. Some of these species have gotten established in parts of Texas and southern United States. A deterrent to the

217

Figure 10.4
Insecticide dustbag.

ability of dung beetles to aid in fly control, however, has been the extensive use of certain dewormers and systemic insecticides in which their residues, when passed out in manure, will often kill dung beetles.

Chemical Control: Chemical control has been the most effective means of reducing horn fly and face fly populations. Historically, horn flies have been the easiest to control because they stay on the animals at all times. Routinely spraying cattle with insecticide sprays can be effective for horn fly control, but usually this is costly and impractical. Most available insecticides available in spray formulations last only 1 to 2 weeks, and cattle need to be handled and brought into a confined area each time they are sprayed. This is not feasible in most management systems. Getting long-lasting insecticide treatments on cattle or getting animals self-treated daily has proven much more effective.

Self-application devices that include insecticide dustbags and backrubbers/ oilers can be effective for both horn flies and face flies (Figure 10.4). These devices need to be in a forced-use treatment situation where the animals make contact with them daily. Placing them in a path to water or salt/mineral source and positioning them so cattle will rub them onto their heads and down their backs will provide the most effective control. The key to their success is having them serviced and in place when the first flies appear in the spring and keeping them serviced throughout the fly season. These devices should not be used with horned cattle, as they tend to destroy or damage them. Insecticides available for use in these devices are certain formulations of pyrethroids and organophosphates.

Figure 10.5
Cow with insecticide ear tags.

Insecticide-impregnated ear tag devices are also available and can provide effective control of horn flies and aid in the control of face flies (Figure 10.5). In these ear tags, the insecticide is in the matrix of the tag. As the insecticide leaches to the surface of the tag, it rubs onto the hair coat of the animal, providing fly control. There are numerous trade names and designs currently marketed containing pyrethroid and organophosphate insecticides. Depending on the product, one or two tags are recommended for installing per animal. For face fly control, it is generally recommended to use two tags per animal, one in each ear. Insecticide ear tag treatments are designed to provide season-long control with a single application.

Ear tags containing pyrethroids have provided excellent control of horn flies and face flies. However, horn fly resistance to these pyrethroids has developed. The organophosphate tags available will control pyrethroid-resistant horn flies. The following management practices should be followed in controlling horn flies and face flies:

1. Install one to two tags per adult animal (as per label instructions) after flies first appear in the spring, preferably in late May to early June. Ear tags release insecticide most efficiently during the first 2 months after application. If applied too early, tags may become ineffective before the end of the fly season.
2. Remove the tags at the end of the fly season (September/October in most areas).
3. If horn fly populations exceed 200 per animal during the fly season, supplement control with organophosphate dustbags or oilers or use feed additives, boluses, or nonpyrethroid pour-on treatments.

219

4. If horn fly resistance is established, remove or don't use pyrethroid tags. Use organophosphate tags or other control practices as described in practice 3 above.

Other methods available for fly control on pastured cattle are the use of insecticide pour-on treatments, oral bolus treatments, or insecticide feed additive/ mineral supplements. With insecticide pour-on treatments, the pour-on product, which is generally formulated as a ready-to-use product, is applied from the poll of the head down the backline to the tailhead using an applicator gun or device (Figure 2.12). Products available as pour-ons include pyrethroids, avermectins, and spinosad. Where pyrethroid-resistant horn flies are present, it is advised to use a non-pyrethroid-resistant pour-on. Generally, these pour-on treatments provide from 4 to 12 weeks of control, depending on the product and environmental conditions.

Bolus treatments are given to animals orally, and the bolus settles in the reticulum where a controlled release of the insecticide passes out in the manure (Figure 2.14). Active ingredients used in bolus formulations are insect growth regulators (IGRs) that prevent immature fly larvae from becoming adults. Another means of oral treatment is by the use of larvicides as feed additives in free-choice loose mineral supplements or solid mineral blocks containing an insecticide (Figure 2.13). These products, when consumed by cattle, pass through the digestive tract of the animal and kill developing fly larvae in the manure or inhibit their development. Products available may contain an IGR (methoprene), spinosad, or tetrachorvinphos. These oral insecticides are not effective against adult flies. For this reason, if cattle located nearby are not receiving an oral larvicide, their manure will successfully breed flies, and this could reduce the benefit of this treatment method.

With any insecticide used, always read and follow label directions for use. For lactating dairy cattle, use only products that are allowed to be used, as dictated on the label.

Flies in Cattle Confinement Production Operations (Beef, Dairy)

Of the flies occurring in cattle confinement operations, including feedlots and dairies, of most concern are stable flies, *Stomoxys calcitrans*, and house flies, *Musca domestica*. Control of these two flies follows similar techniques.

Surveillance/Diagnosis: Stable flies and house flies can cause annoyance to cattle. Stable flies feeding on cattle, especially on the legs, causes foot stomping, tail swishing, animal bunching, and nervousness in cattle (Figure 10.6). Excessive house fly populations can also alter animal behavior. Cattle become reluctant to feed with high numbers of house flies present around feed bunkers, and cattle often bunch together to avoid fly activity. Observing animal behavior can be an indication for the need for fly control.

Figure 10.6
Stable flies on leg of cow. (Photo reproduced with permission by John B. Campbell, University of Nebraska.)

Monitoring fly activity can be accomplished in several ways. The use of light traps, baited fly traps, sticky ribbons, and spot (fly speck) cards are useful in monitoring fly activity, especially house fly activity inside buildings. Electric fly traps, preferably those that either stun flies or simply get them stuck on a sticky pad, can be used in areas accessible to electrical outlets. There are several commercial light traps available that can be used. These should be placed away from competitive light sources and be accessible for changing the sticky pads in them. Flies stuck on sticky pads in these devices can be counted.

Baited fly traps can either be purchased or made. Commercial traps usually consist of a jug design with liquid contents with fly attractant. Traps can be made using gallon plastic milk/food jugs with four 5-cm holes cut in the upper part of the sides (Figure 13.5). Place a small amount of granular fly bait (1 oz) into the jug. The holes allow flies to enter the jug to feed on the bait and die. Baited traps can be suspended with wire from rafters or building support beams.

Sticky fly ribbons can be strategically placed in a facility as well. Place these where fly activity is likely and in a position where they hang vertically with both sides exposed.

Fly spot cards consist of 3" by 5" plain white file cards that can be attached to obvious fly resting surfaces. When flies land on these cards, they usually leave fecal and regurgitation spots (Figure 13.3).

The number of light or baited traps, sticky ribbons, or spot cards needed will vary according to the building size and configuration. Five to 10 units of any of these minimally should be used. These devices should be checked, serviced, or replaced at 7-day intervals and fly activity enumerated. Numbers of flies caught or fly specks counted will vary by facility, but a general rule of thumb is if 250 flies are caught per week per unit, or 100+ fly specks are counted per card, this would normally indicate higher fly activity that needs

Figure 10.7
Olson fly trap.

attention. Fly spot cards can be saved to provide historical documentation of fly activity and control effectiveness.

Outdoors, baited traps can be used as well as sticky traps. Sticky traps can be commercially obtained or made. One type of commercial trap that has shown to be effective in monitoring both house fly and stable fly activity is the Olson fly trap (Figure 10.7). This is a cylindrical trap placed on a wood stake and wrapped with a sticky sleeve (see www.olsonproducts.com). Like indoor monitoring devices, these outdoor traps should be serviced on a weekly basis. Stable flies can be monitored directly by counting flies on the legs of cattle, the preferred feeding location for these biting flies. A minimum of 10 to 15 animals should be included, and flies observed on all four legs counted. Fly control is generally warranted when counts average 5+ flies per leg.

Another aspect of monitoring for fly activity in beef confinement and dairy operations is to identify sites where flies are breeding in and around the operation facilities (Figure 10.8). Designing a map/guide of the operation layout showing buildings, feed bunks, loafing areas, watering sources, hay feeding areas, walkways, and so forth, will be useful in the monitoring activity.

Create a checklist of sites to monitor and keep this as part of a written fly management plan, along with adult fly activity records. This checklist can be used to assess the need for breeding site cleanup and maintenance and for fly control action steps needed.

Figure 10.8
Fly breeding areas around barn.

Potential fly breeding areas to check include the following:

- Areas under fences and perimeter edges and corners of loafing areas
- Under and around feed bunkers and waterers and water tanks
- Old feed in bunkers and spilled along edges
- Moist piles of manure and bedding
- Loading areas for manure hauling and application equipment
- Alleyways and cattle loading areas
- Silage effluent at the base of silage piles or spilled silage
- Around the base of silos
- Around and beneath stored and fed hay bales
- Spilled feed around feed mixing areas and augers
- Calf pens and hutches, particularly in corners and around feed and water sources
- In and around sick animal pens
- In and around maternity pens
- Runoff areas
- Vegetation surrounding pens

Fly Control in Confined Beef and Dairy Operations: Effective control of stable flies and house flies in cattle confinement operations involves an integrated approach. Sanitation is the first and most important step to prevent fly development. No insecticide can be effective for these flies as long as breeding sites exist. The following discussion details the key elements of a total fly pest management program for confined beef and dairy operations, including sanitation and cultural practices, biological control, and the judicious use of insecticides.

Sanitation/Cultural: Proper waste management and sanitation practices are key aspects of an effective fly control program. Beef confinement facilities and dairies should be properly designed or modified to facilitate ease in the cleaning of waste and to minimize accumulations of manure, spilled feed, moisture, and other sources of organic debris.

Beef confinement and dairy operations vary significantly in size and management practices. The following points are recommended sanitation and cultural practices that should be considered for incorporating in an integrated fly pest management program to fit in with any given type of operation:

- Scrape and clean loafing areas frequently.
- Clean stanchions and gutters frequently.
- Clean up spilled feed weekly.
- Divert surface water by grading and providing drainage around barns and facilities.
- Enclose area under feed bunkers and provide concrete aprons around feed bunkers and waterers.
- Maintain working float valves in water tanks to prevent overflows.
- Provide shelters over feed bunkers to prevent excessive moisture from rain and feed wastage.
- Provide curbs and concrete surfaces in loafing areas to aid cleaning.
- Clean out maternity stalls, calf pens, and hutches frequently.
- Move calf hutches at frequent intervals.
- Flush manure pits under slatted floors at frequent intervals to remove solid waste buildup.
- Minimize seepage from silage storage by covering and sealing edges of bagged silage and silage piles.
- Clean around the base of silos and auger runs at frequent intervals.
- Cover and store hay bales on dry land or raised pallets, or store hay in properly designed stacks or hay sheds.
- Screen windows and maintain closed entries into milking parlors and processing areas.
- Keep lagoons free of debris and floating solids and do not overload.
- In feedlots,
    - Maintain mounds to provide a dry area for cattle and drainage for excess moisture to escape.
    - Maintain good slopes in pens to provide good drainage.
    - Maintain drainage systems with enough slope to move the moisture to holding ponds during wet periods.
    - Scrape wet areas to facilitate drying.

Figure 10.9
Weeds at feedlot. (Photo reproduced with permission by John B. Campbell, University of Nebraska.)

- Stock pens with enough cattle to cause effective trampling of wastes.
- Keep weeds and grass cut around the perimeter of pens and buildings to minimize fly resting areas and allow for effective wind movement (Figure 10.9).
- If manure is used as fertilizer and spread directly on farm agricultural fields, spread the manure thin enough for rapid drying.

Biological Control: For beef confinement and dairy operations several insectaries have available parasitic wasps that can aid in fly control (Figure 2.5). Depending on the climate and geographic location, some species of these wasps are more effective than others in reducing fly numbers. The most effective species utilized are in the wasp family Pteromalidae and in the genera *Muscidifurax* and *Spalangia*. These are fly pupal parasites in which the female wasp oviposits one or more eggs inside a fly puparium. The larval wasp kills the developing fly and emerges from the dead fly puparium as an adult wasp. Natural populations of these parasites exist but usually do not develop in high enough numbers to provide noticeable fly control. Commercial release of these fly parasites can aid in the control of flies. It is advised to check with commercial suppliers of these parasites to customize a suitable program geared for a particular operation.

In using these fly parasites for most effective fly control benefit, the following conditions should be met:

225

- Proper sanitation and cultural practices must be carried out. Parasite releases complement these practices but cannot replace them.
- When insecticidal treatment is necessary for supplementing fly control, use insecticides and application methods that are compatible with those for parasites. Nonresidual space sprays and fly baits, if applied properly, usually are safe to use.

Recently a fungus, *Beauveria bassiana*, was registered as a fly control agent. This fungus works by attacking the cuticle of flies, leading to their death. Approved formulations of this fungus are currently being marketed in bait and spray formulations for use in livestock and poultry operations.

Chemical Control: In using chemical insecticides, it is imperative to read and follow all label directions and restrictions. Although some insecticides may be used associated with beef operations, they may have use restrictions with dairy operations.

The following guidelines should be taken into consideration when choosing appropriate insecticides to use:

- Read and follow all label directions for proper mixing instructions, application rates, and precautions.
- Do not use insecticides for dairy cattle or dairy premise use if not labeled for such use.
- Label directions may restrict use to nonlactating dairy cattle.
- Do not contaminate milk, feed, or water when using an insecticide.
- Keep insecticides in original containers, properly labeled, and in locked storage.
- It is illegal to use an insecticide in any manner inconsistent with the label.

The use of insecticides is a viable part of an integrated fly pest management program. They can be used for fly control in several ways. Application methods include space (area) sprays, residual sprays, on-animal sprays, baits, larvicides, and oral treatments.

Space Sprays: Space sprays are designed to provide a quick knockdown of active adult flies in an infested area with no residual activity. They can be applied using various spray equipment that delivers fine mists, aerosols, or fogs (Figure 2.15). They are most feasible in portable equipment that can be walked or driven through a facility or operation (e.g., backpack, cart mounted, or truck mounted). In large outdoor operations, properly equipped aircraft can also be used. Because these sprays do not provide residual control, they should be used when adult flies are most active. Chemicals registered for use include synergized natural pyrethrins, some pyrethroids, and dichlorvos.

Residual Sprays: Residual insecticide sprays are applied to fly resting surfaces such as on walls and support posts. These sprays can be applied with hand sprayers or power equipment (Figure 2.17). Sprays are normally applied onto a suitable surface to the point of runoff. Avoid contamination of feed or water. Classes of chemicals registered for use as residual sprays include some pyrethroids, organophosphates, and spinosad. With formulations available, residual spray treatments will last from 7 to 10 days. The development of genetic fly resistance between any of these classes of insecticides can be deterred by periodic rotation between these different groups of insecticides. If residual sprays are not effective for use in a particular situation, they should not be used.

Animal Sprays: Insecticide sprays on animals in beef confinement and dairy operations can be made, but should only be considered to provide temporary relief of stable fly adults. This is not a feasible approach to control house flies, however. Natural pyrethrins and some pyrethroids are registered for on-animal use.

Fly Baits: Insecticide fly baits are only effective against nonbiting flies, primarily house flies. Blood-feeding flies like stable flies are not attracted to baits. Most bait products available are in granular form and contain such active ingredients as methomyl, imidacloprid, and spinosad (Figure 2.18). Baits should only be applied in locations inaccessible to animals. They can be placed in bait stations, used as scatter baits in infested areas, or mixed with water to form a slurry (if allowed on the label) and applied with a brush on fly-infested surfaces. Baits are best used in buildings and enclosed areas and are less effective in outdoor areas.

Larvicides: A number of insecticides are registered for use as larvicides. They can be used directly on manure and other fly breeding sources (Figure 10.10). This type of application is best utilized when reserved for treatment of fly breeding spots not eliminated by normal sanitation practices. Because of the chemical and physical makeups of manure and other decomposing material, most larvicide treatments break down rapidly.

Oral Larvicides: Oral larvicides are available as controlled-release boluses and feed additives (Figure 2.13). Boluses, when ingested, stay in the reticulum and release insecticide over several weeks which passes out in the manure. With feed additives in mineral supplements or block formulations, the insecticide passes through the digestive system of the animal and is present in the manure to kill fly larvae. Insecticides available include IGRs and tetrachorvinphos. The problem with oral insecticide use is that only the manure from animals given the product is treated. Only house flies breed in manure. Stable flies prefer breeding sources containing decomposing plant material. Also, house flies will breed in other material besides fresh manure. Oral larvicides are most effective where fresh manure is the primary fly breeding source.

Figure 10.10
Spraying manure for fly larval control. (Photo reproduced with permission by John B. Campbell, University of Nebraska.)

MOSQUITOES, TABANIDS, BLACK FLIES, AND BITING MIDGES

Cattle in both confinement and pasture situations can be annoyed and attacked by various blood-feeding Diptera. Mosquitoes, tabanids (including horse flies and deer flies), black flies, and biting midges often are of concern. Each of these groups can cause enough irritation to cattle to sometimes warrant control. In addition, because of the importance of mosquitoes, tabanids, and biting midges as vectors of various animal disease agents, their control takes on added importance.

Surveillance/Diagnosis

Proper detection techniques and surveillance activity are difficult, at best, for mosquitoes, tabanids, black flies, and biting midges on cattle. The following are surveillance guidelines that can be followed associated with each of these groups in and around cattle operations.

Mosquitoes: Assessing adult mosquito activity on cattle is very difficult. Mosquitoes often feed on the underside of animals where there is less hair. Mosquitoes are generally more active in early morning and around dusk. Observing animal behavior during peak mosquito feeding times can sometimes

228

Figure 10.11
CDC mosquito light trap.

indicate a need for control. If animals appear nervous, bunch together, and show excessive tail swishing, this could be an indication of mosquito biting activity during these times.

Mosquito abatement personnel will use various types of mosquito traps to monitor mosquito activity in an area. These often consist of a small light source and a fan to trap mosquitoes into a netted cage. The CDC mosquito light trap is one such device (Figure 10.11). It can be powered with ordinary flashlight batteries. This type of trapping can aid in determining general mosquito activity at a given location. However, there are many variables to take into consideration on how effective traps can be in assessing mosquito incidence on cattle. Not all mosquitoes are attracted to these traps, and trap placement and numbers of traps needed at a location to give accurate population estimates will vary.

Mosquito larval surveillance activity can provide assessments on the extent of mosquito breeding on and around areas where cattle are kept. All mosquitoes require standing water to breed in. The diversity of larval habitats can be difficult to assess, however, especially in pasture situations. Mosquitoes will be found breeding in such habitats as tree holes, ditches, and other low-lying spots that accumulate water; water troughs and tanks; discarded tires; in trash piles where there are open containers; isolated pockets of standing water associated with small water streams; and numerous other sites. In confinement areas, mosquitoes can be found breeding in waterers and water troughs, runoff areas of outdoor lots, and anaerobic waste lagoons, among other locations. Also, tires used to hold silage coverings in place have resulted in mosquito breeding activity. Inspections of any of these locations will help to assess where mosquito breeding may be occurring. Some mosquitoes, however, such as certain *Aedes* sp., will fly for several miles seeking a blood meal and may not be found breeding on the property.

Mosquito larval sampling can be done using a white enamel or plastic dipper attached to a handle (Figure 3.4). This is used to dip into the water to catch mosquito larvae. Because most mosquito larvae are easily disturbed, some practice is needed to effectively use this sampling technique. By identifying mosquito breeding sites, they can be possibly eliminated, altered, or chemically treated to reduce actual mosquito breeding that may be occurring on a property, realizing, however, that some mosquito breeding activity may be occurring elsewhere.

Tabanids (Horse Flies and Deer Flies): Tabanids can certainly be of importance to cattle well-being and productivity, not only from annoyance and irritation of their biting activity, but also because of disease concerns (e.g., anaplasmosis). Tabanids are typically strong fliers and are capable of flying several miles from their breeding habitats. Because they breed in a variety of marshy and wetland areas, this makes surveillance a difficult task. Observing tabanid activity on animals and observing animal behavior are probably the best measures of tabanid abundance and annoyance. Certain adult trapping methods have been developed which can aid in assessing tabanid activity. One such trap is the Epps biting fly trap (Figure 10.12). This trap is equipped with contrasting surface areas and a transparent area that attracts the flies, which then are knocked down into a soapy water tray. Properly placed, this kind of trap can also aid in tabanid fly control.

Black Flies: Black flies are also capable of flying long distances, but cattle maintained near unpolluted streams and rivers where black flies breed are most frequently attacked. Surveillance of larval breeding sites is not practical. On animals, black flies will often be found in the ears. They will also feed on the underside on the belly and axillaries. Heavily infested cattle will shake their heads and often bunch together to avoid attack. Black fly activity is usually more prevalent in the late spring and early summer.

Biting Midges: Detection of these small flies is very difficult. They usually go unnoticed when feeding. Larvae breed in a wide variety of aquatic to semiaquatic habitats. In confinement operations, some species of *Culicoides*, especially *C. variipennis*, can sometimes be found breeding in feedlot runoff areas (Figure 10.13). Larvae can be sampled for at the muddy edge of water. Taking a small sample of submerged mud at the water edge and placing it in a white enamel pan is a method used to look for the presence of *Culicoides*

Figure 10.12
Epps biting fly trap.

Figure 10.13
Culicoides *breeding site—feedlot runoff area. (Photo reproduced with permission by John B. Campbell, University of Nebraska.)*

larvae. Upon close examination, the small larvae can be seen by their unique serpentine movement in the pan. With the more frequent occurrence of epizootic hemorrhagic disease (EHD) virus transmitted by *Culicoides* in cattle, surveillance in stagnant water breeding sources around pastures and confinement areas can help to pinpoint possible sources where the *Culicoides* midges are breeding.

Control

Control of these biting flies is difficult to accomplish on cattle. On pastured cattle, control is impractical. Insecticide-impregnated ear tags can provide some relief of black fly feeding activity in the ears but are not effective for tabanids, mosquitoes, or biting midges. For confinement beef and dairy cattle, whole-animal sprays and mists can provide temporary relief, but retreatment is frequently needed.

When breeding habitats can be identified, efforts can be made to eliminate or reduce breeding activity. Filling in low-lying spots where water accumulates, minimizing water leaks at waterers, and preventing stagnant water buildup can help. Properly functioning holding ponds and anaerobic waste lagoons with minimal surface solids and vegetation-free shorelines will also help to minimize mosquito and biting midge breeding. Supplemental insecticide application to standing water areas not easily eliminated may also help when needed.

Control efforts for tabanids in marshes and wetlands are impractical. Also, it is illegal in many areas to destroy natural wetlands.

Although some efforts have been employed to control black fly larvae in streams using certain organophosphate insecticides and the bacterium *Bacillus thuringiensis,* this has not proven to be cost effective for area-wide control effectiveness.

SCREWWORMS/BLOW FLIES

The success of the primary screwworm, *Cochliomyia hominvorax*, eradication program has resulted in complete elimination of this pest in the United States and Mexico. Several other species of calliphorid blow flies can infest wounds of cattle, however, but only feed on dead tissue. Detection of blow fly activity on cattle is a matter of looking for and examining wounds on animals for the presence of blow fly larvae (Figure 10.14). When infestations are found, there are several insecticide products available that can be applied directly onto the wound to kill larvae. With extensive wounds, it is advised to seek veterinary assistance.

The best deterrent to preventing blow fly strike on animals is to follow management practices aimed at reducing the frequency of wounds on animals.

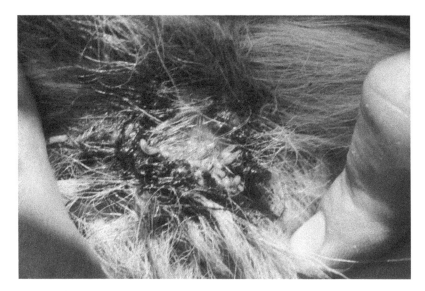

Figure 10.14
Blow fly infestation on cow.

Proper timing of castration, dehorning, and branding during times or seasons when blow flies are not active is preferred. Whenever these activities are performed, a preventive application of an improved insecticide spray or smear on the wound will help prevent blow fly strike.

CATTLE GRUBS

Surveillance/Diagnosis

During egglaying activity of adult female cattle grubs, known as heel flies and bomb flies, the behavior of cattle can indicate the presence of these flies. In response to the buzzing of adult female flies, cattle will run about wildly with their tails straight up. This behavior called *gadding* is a sign that cattle grubs are present. Adult egglaying usually ends in late summer and fall. Once emerged, cattle grub larvae penetrate into an animal, and they take from 9 to 10 months to migrate through the body of their host before reaching the back.

The presence of cattle grubs is best examined for in late winter to early spring after they have formed warbles in the back of a cow. Warbles can be detected by palpating (rubbing) the cow's backline from the shoulder to the tail, feeling for the raised bumps. By parting the hair around a warble, the breathing hole of the grub may be visible (Figure 4.5). Five or fewer grubs per back are generally considered as a light infestation, and 15 or more are considered a heavy infestation. The presence of grubs in the back can indicate a need for control in the late summer or autumn.

Control

If cattle are detected to have cattle grubs in a herd by spring palpation or if there has been a history of cattle grub activity, then control efforts can be made. Depending on geographical location, insecticide treatments should be applied in late summer to early winter. Applications should be made after egg-laying activity ceases, but before grubs reach the esophagus region or back region. Cattle grubs killed in the esophagus region can cause adverse reactions from swelling. If grubs are killed when they reach the back area, this can cause paralysis problems in the animal from enzymes released by the dead, decomposing grub affecting the central nervous system.

Cattle grubs can be easily controlled with systemic insecticide and parasiticide applications. Registered organophosphate, avermectin, moxidectin, and eprinomectin products are available as sprays, pour-ons, spot-ons, or injections (Figure 2.12). These systemic treatments are absorbed into the tissues where they kill the migrating early instar cattle grub larvae. These treatments also aid in the early season control of blood sucking cattle lice. Organophosphate and avermectin systemic treatments can only be used on beef cattle and non-lactating dairy cows and heifers. They cannot be applied to lactating animals because of the possibility of insecticide residues appearing in milk. Also, these products have slaughter restriction intervals after treatment. For specific interval times, check the product label. For lactating dairy cattle, pour-on treatments containing either moxidectin or eprinomectin can be used with no slaughter or milking restrictions.

The use of systemic insecticide treatments has proven to provide up to 100% control of cattle grubs. The use of systemic insecticides over several years has dramatically decreased the incidence of cattle grubs in North America and has resulted in near elimination of these pests in several areas. Compulsory use of systemic insecticides over a 2- to 3-year period in a given area, through cooperative efforts of area cattle producers, could eliminate these parasites in such areas.

LICE

Lice on cattle are of most concern during the colder months in North America. Detection and control of lice are directly on animals because they are continuous obligatory parasites, spending their entire lifecycle on their host.

Surveillance/Diagnosis

Lice infestations often go unnoticed, especially in lower infestations. It is not until cattle show observable hair loss and excessive scratching behavior that cattle producers recognize that there is a lice problem (Figure 10.15). At this time, lice populations have increased above economic injury levels. Cattle

Figure 10.15
Lice-infested cattle scratching.

should be examined for lice infestations before populations have reached harder-to-treat higher infestations. Surveillance activity should start in late summer or early autumn. Sampling involves getting animals in a squeeze chute and carefully inspecting sections of skin on a representative number of animals in a herd (10 to 15 animals). Five to six areas on the body should be examined, including the head, neck, shoulders, back, hips, and tail. The presence of lice should be made by parting the hair and looking for lice on the skin. If lice counts per hair part or per square inch average 10 or more lice, then treatment is generally recommended. Determination should also be made if the lice are biting lice, *Bovicola bovis*, or one or more of the blood sucking lice. This could determine how control should be made. Careful and regular monitoring for lice can detect problems before lice infestations get out of control.

Control

Cattle maintained on a high plane of nutrition in beef feeding and dairy operations generally do not carry heavy lice infestations. In feedlots and dairies, new animals brought into a herd should be isolated and carefully examined for lice. If found, they should be treated before they are introduced in with the rest of the herd. In dairy operations, housing calves in separate hutches will reduce infestations on these replacement animals.

There are several insecticides and application methods available for lice control on cattle. Methods available for control include sprays, dust, pour-ons, injections, and self-application devices (e.g., dustbags, oilers, backrubbers). Some insecticides available for lice control may have slaughter interval restrictions and/or restrictions for lactating dairy cattle. As with any insecticide application, read and follow all label directions before using any insecticide. The choice for application method and insecticide used should be based on meeting individual needs and production management practices.

Whole-animal sprays can be used. Sprays should be made using pressure sprayers of 40 lb psi or less. Spraying cattle for lice should be made on warmer, dry days. Spraying cattle in colder weather can result in cold stress to the animals. Insecticides available for spray treatments include organophosphates, pyrethroids, and amitraz. Dusts can also be used by direct application on animals. Pyrethroid and tetrachorvinphos dusts are available and can be used on both beef and lactating dairy cattle. Pour-on treatments can be applied to animals by dispensing the product on the backline of the animal. Insecticides and parasiticides available in pour-on products include organophosphates, pyrethroids, avermectins, moxidectin, and eprinomectin. Organophosphate products and avermectins can only be used on beef cattle and nonlactating dairy cattle and heifers; pyrethroids, moxidectin, and eprinomectin can be used on lactating dairy cattle.

Because louse eggs are resistant to most insecticides, animals should be treated twice with sprays, dusts, and some pour-ons, at 8- to 10-day intervals. The first application will kill active adults and nymphs, and the second treatment will kill nymphs recently emerged from eggs and any not killed by the first treatment. Cattle with a history of louse infestations should be treated annually in the autumn or early winter before lice populations increase. If animals are treated for cattle grub control, this will provide early season lice control. However, retreatment may be necessary later in the winter if lice infestations increase. If there is a history of cattle grub infestations in a herd and cattle grubs are not treated for, do not use a cattle grub product for lice control after recommended cattle grub treatment cut-off dates for a given geographic area. Also, biting (or chewing) lice are not readily controlled with systemic treatments. Surface-acting insecticide treatments need to be used when biting lice are a problem.

The use of self-application treatments can be an effective means of lice control. There is an array of devices that can be employed including dustbags, backrubbers, and oilers. The key to their effectiveness is to place them in areas where animals will contact them frequently enough to treat themselves. Infested animals will use these devices to rub and scratch up against and get a small dose of insecticide (Figure 2.11). It is important to keep these devices serviced regularly with adequate amounts of insecticide and assure they are in working condition. Depending on the number of animals in a herd, multiple self-application devices may be needed to adequately treat a herd or group of animals.

TICKS

Tick infestations on pastured cattle will vary by geographical areas around the United States and by season of the year. Also, some species will attach over the whole body (e.g., *Amblyomma americanum*) (Figure 10.16), whereas others will be confined to certain body regions, such as *A. maculatum* that will be found primarily in the ears of cattle (Figure 10.17). One-host ticks will spend most of their lifecycle on their host, and some three-host ticks feed only as adults or spend much less time on their host. Surveillance and control approaches will differ depending on the tick species targeted.

Surveillance/Diagnosis

Direct examination of animals is often the most effective means of assessing tick abundance. To do so, adequate facilities are needed to restrain cattle for thorough examination. In examining animals for ticks that feed on several body regions, it is recommended to check specific areas such as the head, neck, brisket, dewlap, escutcheon, axillary regions, and perineum for attached ticks. Cattle ears should be examined when *A. maculatum* and spinose ear tick, *Otobius megnini*, infestations are of concern.

Figure 10.16
Lone star tick infestation on deer ears.

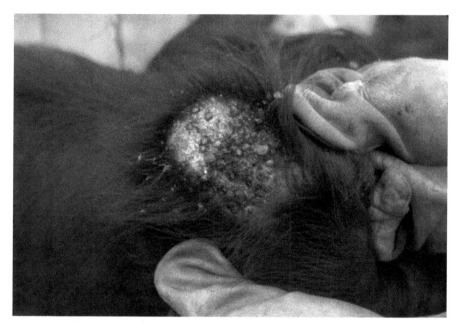

Figure 10.17
Gulf Coast tick infestation.

With three-host ticks, surveillance can also be made of the free-living stages. A common method employed is "dragging." This involves dragging a cloth on the ground and vegetation in suspected tick-infested habitats. Ticks will cling to the cloth and can be counted, collected, and identified. A suggested drag cloth is a sheet of muslin 4 ft long by 3 ft wide attached to a 4 ft wood pole at one end. A rope attached to each end of the pole can be used to drag the cloth. Wild animal trapping and examining natural hosts for ticks can also be done. Several three-host ticks utilize ground-nesting birds and rodents for larval and nymphal stages.

Control

Tick control primarily involves treatment of animals. However, habitat management can aid in reducing tick abundance in an area. Habitat modifications, such as burning or clearing vegetation, can make a habitat unsuitable for tick survival and reduce harborage for non-cattle hosts of ticks. Selective grazing and pasture rotations often reduce cattle exposure to tick populations. By keeping animals out of certain pastures for a given season, the number of fed ticks that would serve to build-up a population is reduced. This has often been referred to as pasture "spelling."

In the south and southwestern United States where cattle tick infestations can be heavy, tick-resistant breeds of cattle (e.g., *Bos indicus*) often perform

Figure 10.18
Cattle-resistant breed research: Bos indicus/B. taurus.

much better than European cattle breeds (*Bos taurus*) (Figure 10.18). Several newer breeds of cattle with *Bos indicus* cross-breeding have done quite well in areas with heavy tick infestation.

Control of ticks with acaricides can either be directed against the free-living stages of the host in the environment or against the parasitic stages on the host. Using acaricides off the host in tick-infested areas is of limited value. Residual acaricide treatment can be applied on infested areas. This type of application is best utilized as limited border treatment at edges of small pastures and along edges of wooded areas. Pesticides available are certain organophosphates, carbamates, and pyrethroids. Habitat acaricide treatment is not practical for whole-pasture treatment.

Host treatment is the most common approach for controlling ticks associated with cattle. Acaricides are applied on cattle by various means, including animal sprays, whole-body dipping, acaricide-impregnated ear tags, pour-on treatments, and injectable treatments. In using sprays, a high-pressure spray that wets the entire animal is recommended (Figure 10.19). Dipping vats have been used extensively for tick control on cattle (Figure 2.8). With dipping vats, cattle are run through the vat full of acaricide-treated solution to entirely expose the animal from head to toe. When the animal leaves the vat, excess solution drains from the animal back into the vat. Dipping vats need to be maintained with proper solution levels and percentage of active ingredient when used for several animals over extended periods of times. Retreatment

Figure 10.19
Spraying cattle with insecticide.

may be needed several times during a season when using sprays and dips. Pour-on applications using such active ingredients as pyrethroids, avermectins, moxidectin, and eprinomectin have been very effective for tick control. Injectable treatments using avermectins have also been shown to be effective. Acaricide-impregnated ear tags (Figure 2.9) have provided season-long control of ticks, especially for ticks such as *A. maculatum* that primarily attaches in the ears.

MITES

Mites found on cattle are continuous obligatory parasites that do not survive very long away from their host. Cattle mite infestations, especially sarcoptic mange and psoroptic scabies, are highly infectious. The severity of sarcoptic mange and psoroptic scabies requires infested animals to be quarantined and treated.

Surveillance/Diagnosis

An indication of mite infestation on cattle is visible lesion development and poor condition in cattle. To confirm infestation, however, skin scrapings need to be taken at infested sites on animals. With *Psoroptes ovis*, that causes cattle scabies, the lesions usually have the appearance of a dry scab surrounded by successive zones of moist crust and reddened skin (Figure 7.8). The mites are most active at the edge of the scab and are found on the skin surface. Using a scalpel and scraping along the surface of the edge of the scab can be done to confirm infestation. The scraped material should be examined using a microscope to confirm the presence of mites and to aid in species identification. *Chorioptes bovis* infestations on cattle can be sampled the same way because they also occur on the surface of the skin. With *Sarcoptes scabiei*, infested areas are characterized by hair loss and heavy, thick crusts or scabs. These mites feed on the moist layers under the scabs. Scrapings for sarcoptic mites need to be deep enough under the crusty scabs to expose the moist layers of skin. Scraping cattle for mites is best done by a veterinarian or other trained professional.

Control

Confirmation of sarcoptic and psoroptic mite infestations requires animals to be quarantined and proper control measures to be followed. These are federally reportable diseases in cattle. Chorioptic scabies is considered a reportable disease in some states. Reporting sarcoptic mange and psoroptic scabies is mandatory, and regulations specify the chemicals and methods of application to be used. Approved methods of application for beef cattle and nonlactating dairy cattle are by total-body dipping or spray-dip machines using approved acaricides (e.g., amitraz, coumaphos, permethrin, phosmet), or by injection using ivermectin. For lactating dairy cattle, animal sprays using approved acaricides (e.g., permethrin, coumaphos) and pour-on applications (e.g., moxidectin, eprinomectin) can be used. Registered acaricides available for use may vary from state to state.

Several management practices can be followed to prevent infestations:

- Be cautious when purchasing or boarding new animals.
- Avoid animals that show visible skin lesions or appear itchy or agitated.
- Isolate newly purchased animals from the rest of the herd for several weeks.
- In feedlots, use preventive chemical treatments on new animals brought in.
- Clean stalls used to hold new or infested cattle, and add fresh bedding before using.
- Disinfect grooming tools and other instruments used on animals.
- Keep animals well nourished (cattle in poor condition are more susceptible to infestation than healthy, well-fed animals).
- If any animals show signs of itchiness or characteristic lesion development, call a veterinarian to check the herd.

FIRE ANTS

Imported fire ants are found in many areas spread across the southern United States and are quite common in pastured cattle operations (Figure 9.15). Fire ant activity can cause some economic consequences to cattle production from injury or death of cattle, forage and hay infestations, and to worker health and well-being. Control considerations should be based on the extent that fire ant activity affects cattle production practices.

The broadcast application of registered fire ant bait products is often the most cost-effective way to reduce fire ants in large areas of land. These treatments are most suitable where fire ant mounds are numerous (20 or more mounds per acre). Where fire ant mounds are fewer than 20 per acre or there is only a small area to treat, individual mound treatments can be made. Various products such as baits, mound drenches, or sprays are available containing insecticides such as insect growth regulators (IGRs), pyrethroids, and carbamates. Check state registrations for individual products approved for use.

CHAPTER 11

MANAGEMENT OF SWINE PESTS

TABLE OF CONTENTS

According to the U.S. Department of Agriculture (USDA), there are over 65,000 hog producers in the United States with a swine inventory of over 67 million swine. The five top hog-producing states (Iowa, North Carolina, Minnesota, Illinois, Indiana) represent nearly 70% of the U.S. hog inventory. The annual U.S. value of production based on 2007 statistics was nearly $13.5 billion.

Arthropod pests cause North American swine producers millions of dollars annually due to performance loss in reduced weight gains, reduced reproductive capacity, less feed efficiency, veterinary costs, direct physical harm or damage caused by arthropod attack, and the costs of control in reducing pest population levels. Hog lice, *Haematopinus suis,* and sarcoptic mange mites, *Sarcoptes scabiei,* account for a majority of this expense (more than 70%), with flies and other insect problems accounting for the rest. In this chapter, discussion will be made on pest management strategies in controlling the various arthropods that affect swine production.

HOG LICE AND MITES

The two primary external parasites of economic importance associated with swine are the hog louse and sarcoptic mange mite. The hog follicle mite, *Demodex phylloides*, sometimes is found but generally is not of economic concern.

Surveillance/Diagnosis

Hog lice, reaching a length of 6.5 mm, are the largest blood-sucking lice infesting domestic animals (Figure 11.1). Consequently, they can be readily seen upon visual examination. The best means for assessing lice infestations is by making total body counts on individual animals. This can be done by one person constraining a pig using a cable hog noose and another person examining the animal for adult lice, nymphs, and eggs (attached to hair). Particular attention should be made behind the ears, in folds of skin around the neck, and in tender areas of skin at the axillaries and belly area. A minimum of 10 to 15 animals in a pen or group should be examined. If established lice infestations are present, control efforts should be considered.

Sarcoptes scabiei mites burrow through the epidermal skin layers, making extensive tunnels where adult female mites lay their eggs (Figure 7.10). This tunneling causes severe pruritis and the resultant formation of cutaneous lesions. Proper sampling techniques need to be followed to locate and collect these mites under the crusty lesions in the moist tissue underneath. Proper identification of sarcoptic mange is important because several conditions can contribute to dermatitis in swine. The following information describes proper sarcoptic mange sampling techniques:

1. Materials for proper sampling for mange mites: bone curette (#3 to #6), wood applicator sticks, standard clean microscope slides, slide tray or

Figure 11.1
Hog lice infestation.

holder, and mineral oil. Also have a cloth or tissues available for cleaning the bone curette between use on different pigs.
2. If the animal is infested, mites will typically be found on the inside
   of the ear. However, samples can be obtained at other locations
   on the animal where lesions appear. Dispense mineral oil in the
   cavity of the bone curette. At a suitable site where suspect lesions
   are found, firmly scrape the skin to obtain enough tissue-lesion
   material to fill the cavity of the bone curette (Figure 11.2). It is
   essential to obtain sample material under the scab or lesion where
   moist living tissue is found. This is where the mites will be found.
   A superficial sample of dead skin or tissue will rarely have mites.
   In the ears, samples are best obtained by scraping between the ear
   pinnae at the base of the ears.
3. Onto a clean microscope slide, dispense a small single drop of mineral
   oil. Remove the contents of the scraped material in the cavity of the
   bone curette onto the slide using a wood applicator stick. Uniformly
   crush and spread the scraped material in the mineral oil on the slide
   surface. Be careful not to have too much material on the slide that
   cannot be easily examined. Mark each slide with a marker as to date,
   animal identification, and farm location.
4. The scraping, once properly dispensed on the slides, can be examined up to 24 hours after the scraping is made. The material should
   be examined under low magnification of a compound microscope.

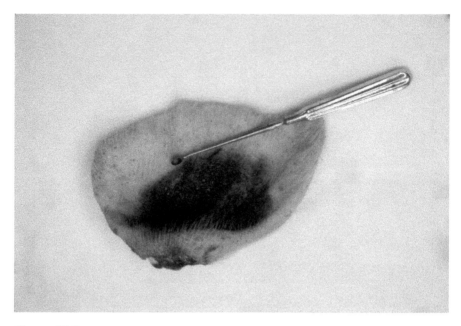

Figure 11.2
Sarcoptic mange mite sampling with a bone curette.

Any adult or immature mites present will often be seen still moving. Also look for any small, coccoid mite eggs that may be present (Figure 11.3).

Control

Good management practices are a must if lice and mange on hogs are to be controlled. These parasites are introduced onto a farm by the introduction of new animals. Any external parasite program should start with good biosecurity. All new animals brought into a herd should be isolated in a separate building until they can be treated for internal and external parasites as well as other disease problems.

The initiation of a total herd control program can be very effective when followed strictly and continually. Such a program is composed of an initial treatment followed by a maintenance schedule. Initially, all animals in the herd should be properly treated. If sprays are applied, a second treatment should be made 5 to 21 days following the first treatment for complete control of newly emerged lice or mites. Also, treat all floors, walls, and equipment surfaces that come in contact with the animals. Before treatment, be sure to remove or empty all feeders and waterers from pens if animals are to be sprayed there.

After initial treatments, all sows should be routinely treated prior to farrowing. Treat piglets at weaning. Treat boars every 3 months. Isolate any new animals and treat before they are comingled with the rest of the herd.

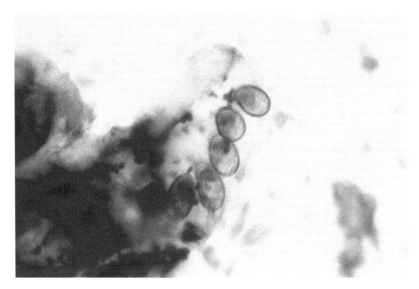

Figure 11.3
Sarcoptic mange mite eggs.

Choosing the right external parasite control is important. Some products available will kill hog lice but are ineffective against sarcoptic mange mites. Products that will kill mites will also kill lice. Sarcoptic mange is generally more difficult to control. Choose a product and application method that address the existing problem and swine production system. Various insecticides/acaricides are available for hog lice and mange control as sprays, dips, oilers, dusts, pour-ons, injectables, or bedding treatments. Available products include such active ingredients as amitraz, coumaphos, malathion, methoxychlor, permethrin, tetrachlorovinphos, and ivermectin.

Sprays are most effective when applied with a power sprayer or other equipment large enough to wet the animals thoroughly, using 2 to 4 quarts of finished spray per hog. It is essential to treat the entire animal, including the inside of the ears, the underside between the legs, and skin-fold areas. Sprays should only be used on warm and sunny days when the animals can dry rapidly. In larger herds, dip-vats can be employed, not only to help cut labor costs but to obtain total skin penetration. Dusts are less effective but can be used if only a few animals are to be treated. Pour-ons and injectable treatments are also available for use. Injectable applications containing ivermectin have been shown to be very effective in controlling both of these pests as well as providing control of selected internal parasites.

Always read and follow label directions of any product used to ensure safe and effective treatment. Withdrawal periods, as specified on the labels, must be carefully observed because of the residue-producing potentials of some of these chemicals. Read the label information for withdrawal times, proper product usage, and application rates. Do not overtreat animals with any compound.

FLIES

Fly control is an integral part of a swine pest management program. Flies associated with swine operations not only add to environmental contamination, but they can also hamper swine production by their blood-feeding activities or annoyance to the animals, and they can potentially carry and transmit pathogenic organisms.

Several species of biting flies may be present around swine and associated facilities. Included among them are stable flies, *Stomoxys calcitrans*, mosquitoes, horse and deer flies, and biting gnats, including black flies and biting midges. Of nonbiting flies, the house fly, *Musca domestica*, and other filth-breeding flies can be a problem. Also, moth flies (family Psychodidae) can be a problem in operations that incorporate liquid manure pits.

Surveillance/Diagnosis

Monitoring fly activity can be accomplished in several ways. The use of light traps, baited fly traps, sticky ribbons, and spot (fly speck) cards are useful in monitoring fly activity, especially house flies inside buildings. Electric fly traps, preferably those that either stun flies or simply get them stuck on a sticky pad, can be used in areas accessible to electrical outlets. There are several commercial light traps available that can be used. These should be placed away from competitive light sources and should be accessible for changing the sticky pads in them. Flies stuck on sticky pads in these devices can be counted.

Baited fly traps can be either purchased or made. Commercial traps usually consist of a jug design with liquid contents with fly attractant. Traps can be made using gallon plastic milk/food jugs with four 5-cm holes cut in the upper part of the sides (Figure 13.5). Place a small amount of granular fly bait (1 oz) into the jug. The holes allow flies to enter the jug to feed on the bait and die. Baited traps can be suspended with wire from rafters or building support beams.

Sticky fly ribbons can be strategically placed in a facility as well. Place these where fly activity is likely and in a position where they hang vertically with both sides exposed.

Fly spot cards consist of 3″ by 5″ plain white file cards that can be attached to obvious fly resting surfaces. When flies land on these cards, they usually leave fecal and regurgitation spots (Figure 13.3).

The number of light or baited traps, sticky ribbons, or spot cards needed will vary according to the building size and configuration. Five to 10 units of any of these minimally should be used. These devices should be checked, serviced, or replaced at 7-day intervals and fly activity enumerated. Numbers of flies caught or fly specks counted will vary by facility, but a general rule of thumb is if 250 flies are caught per week per unit, or 100+ fly specks are counted per card, this would normally indicate higher fly activity that needs attention. Fly spot cards can be saved to provide historical documentation of fly activity and control effectiveness.

Outdoors, baited traps can be used as well as sticky traps. Sticky traps can be commercially obtained or made. One type of commercial trap that has shown to be effective in monitoring both house fly and stable fly activity is the Olson fly trap. This is a cylindrical trap placed on a wood stake and wrapped with a sticky sleeve (see www.olsonproducts.com). Like indoor monitoring devices, these outdoor traps should be serviced on a weekly basis (Figure 10.7).

Another aspect of monitoring for fly activity in swine operations is to identify sites where flies are breeding in and around the operation facilities. Designing a map or guide of the operation layout showing buildings, feeders, loafing areas, watering sources, walkways, and so forth will be useful in the monitoring activity.

Create a checklist of sites to monitor, and keep this as part of a written fly management plan, along with adult fly activity records. This checklist can be used to assess the need for breeding site cleanup and maintenance and for fly control action steps needed.

Control

Sanitation/Cultural   Flies develop in moist manure and wet, decaying organic matter of all types found around swine operations. A thorough sanitation program is a must to minimize fly populations in and around swine facilities. All manure, spilled feed, wet straw, and decaying plant material should be removed weekly to break the breeding cycle of flies. This can be accomplished by spreading manure and other waste material to dry or by placing it in pits or lagoons to liquefy. If a liquid manure pit is utilized, do not allow accumulations of manure above the waterline, either floating or sticking to the sides, because this is an ideal site for muscoid fly and moth fly production.

For mosquitoes, elimination or alteration of mosquito-breeding areas is often accomplished with good management practices. Draining or emptying containers and filling in low areas where standing water accumulates will help. In liquid waste lagoons, efforts can be made to not allow accumulations of manure or other solid waste above the waterline, and to eliminate submergent shoreline vegetation to reduce mosquito breeding. Also, keeping weeds and high grass cut down will reduce outside mosquito and fly harborage sites. For reducing fly, mosquito, and midge biting activity on pigs kept in outdoor pens, installing a fan to blow over the animals will reduce biting activity.

Biological Control   For swine operations, several insectaries have available parasitic wasps that can aid in fly control. Depending on the climate and geographic location, some species of these wasps are more effective than others in reducing fly numbers. The most effective species utilized are in the wasp family Pteromalidae and in the genera *Muscidifurax* and *Spalangia* (Figure 2.5). These are fly pupal parasites in which the female wasp oviposits one or more eggs inside a fly puparium. The larval wasp kills the developing fly and emerges from the dead fly puparium as an adult wasp. Natural populations of these parasites exist but usually do not develop in high enough numbers to provide noticeable fly control. Commercial release of these fly parasites can aid in the control of flies. It is advised to check with commercial suppliers of these parasites to customize a suitable program geared for a particular operation.

In using these fly parasites, for most effective fly control benefit the following conditions should be met:

- Proper sanitation and cultural practices must be carried out. Parasite releases complement these practices but cannot replace them.
- When insecticidal treatment is necessary for supplementing fly control, only use insecticides and application methods compatible with parasite use. Nonresidual space sprays and fly baits, if applied properly, usually are safe to use.

Research studies have looked at other candidate biological control agents. Recently, a fungus, *Beauveria bassiana*, has shown promise as a fly control agent. This fungus works by attacking the cuticle of flies, leading to their death. Approved formulations of this fungus are currently being marketed for use in livestock and poultry operations.

Chemical Control   The use of insecticides is a viable part of an integrated fly pest management program. They can be used for fly control in several ways. Application methods include space (area) sprays, residual sprays, baits, on-animal sprays, baits, larvicides, and oral treatments.

Space Sprays: Space sprays are designed to provide a quick knockdown of active adult flies, mosquitoes, biting midges, tabanids, and moth flies in an infested area with no residual activity. They can be applied using various spray equipment that delivers fine mists, aerosols, or fogs. They are most feasible in portable equipment that can be walked or driven through a facility or operation (e.g., backpack, cart mounted, or truck mounted) (Figure 2.15). Because these sprays do not provide residual control, they should be used when adult flies are most active. Chemicals registered for use include synergized natural pyrethrins, some pyrethroids, and dichlorvos.

Residual Sprays: Residual insecticide sprays are applied to fly resting surfaces such as on walls and support posts. These sprays can be applied with hand sprayers or power equipment (Figure 11.4). Sprays are normally applied onto a suitable surface to the point of runoff. Avoid contamination of feed or water. Classes of chemicals registered for use as residual sprays include some

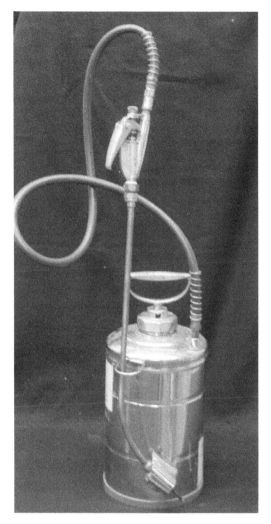

Figure 11.4
Hand sprayer for residual spray application. (Photo courtesy of Patrick Jones, Purdue University.)

pyrethroids, organophosphates, and spinosad. Residual spray treatments will typically last from 7 to 10 days. The development of genetic fly resistance between any of the classes of insecticides can be deterred by periodic rotation between these different groups of insecticides. If residual sprays are not effective for use in a particular situation, they should not be used.

Animal Sprays: Insecticide sprays on animals can be made but should only be considered to provide temporary relief of stable fly adults that may occasionally bother hogs. This is not a feasible approach to control house flies, however. Natural pyrethrins and some pyrethroids are registered for on-animal use.

Fly Baits: Insecticide fly baits are only effective against nonbiting flies, primarily house flies. Blood-feeding flies like stable flies are not attracted to baits. Most bait products available are in granular form and contain such active ingredients as methomyl, imidacloprid, and spinosad (Figure 2.18). Baits should only be applied in locations inaccessible to animals. They can be placed in bait stations, used as scatter baits in infested areas, or mixed with water to form a slurry (if allowed on the label) and applied with a brush on fly-infested surfaces. Baits are best used in buildings and enclosed areas and are less effective in outdoor areas.

Larvicides: A number of insecticides are registered for use as larvicides. They can be used directly on manure and other fly breeding sources. This type of application is best utilized when reserved for treatment of fly breeding spots not eliminated by normal sanitation practices. Because of the chemical and physical makeups of manure and other decomposing material, most larvicide treatments break down rapidly.

Oral Larvicides: Oral larvicides are available as feed additives. Insecticides available include tetrachlorovinphos. The problem with oral insecticide use is that only the manure from animals given the product is treated. Only house flies breed in manure. Stable flies prefer breeding sources containing decomposing plant material. Also, house flies will breed in other material besides fresh manure. Oral larvicides are most effective where fresh manure is the primary fly breeding source.

FLEAS

Hog operations sometimes become infested with fleas, especially cat fleas, *Ctenocephalides felis*. Cat fleas can become an issue with pigs kept on soil lots or on straw or other suitable bedding in barns. Pigs and farm workers can be bitten. Often, fleas build up in such operations because cats are allowed to run freely throughout a hog facility.

Control of flea infestations begins by removing cats or dogs and keeping them out of swine facilities. These animals have no purpose in the hog operation. They can actually pose a biosecurity hazard. Also, cats do not effectively control rodents that may be of concern. If hogs are being infested, infested pigs should be sprayed individually with an approved insecticide. Several pyrethroid and organophosphate products are available for use on pigs and will kill adult fleas. Also, insecticide sprays should be applied on the floor and walls of barns or pens after removal of old bedding and on the soil in infested lots to kill and eliminate immature stages of fleas. Swine housing should be rodent-proofed as best as possible, and a rodent control program should be established to minimize rats and mice as flea hosts.

CHAPTER 12

MANAGEMENT OF SHEEP AND GOAT PESTS

TABLE OF CONTENTS

According to the U.S. Department of Agriculture (USDA), there are over 70,500 sheep producers in the United States with a total inventory of over 6 million sheep and lambs. The annual U.S. value of production (sheep and wool) based on 2008 statistics was over $4 billion. There are also over 108,000 goat producers in the United States with over 3 million goats, including Angora, milk, and other goats.

Arthropod pests cost North American sheep and goat producers millions of dollars annually due to performance loss in reduced weight gains, reduced reproductive capacity, less feed efficiency, wool loss, reduced milk production, veterinary costs, direct physical harm or damage caused by arthropod attack, and the costs of control in reducing pest population levels. Sheep keds, *Melophagus ovinus*, the sheep bot fly, *Oestrus ovis*, and lice account for a large percentage of the losses incurred. Other pests, including wool maggots, other Diptera, mites, and ticks, account for the remainder of losses. In this chapter, discussion will be made of pest management strategies in controlling the various arthropods that affect sheep and goat production.

SHEEP KEDS, MELOPHAGUS OVINUS *(LINNAEUS)*

Surveillance/Diagnosis

Sheep keds are continuous obligatory external parasites spending their entire lifecycle in the fleece of their host (Figure 12.1). They build up in higher numbers during cold weather. They are most commonly found on the neck, shoulders, breast, flank, rump, and on the underside of animals around the stomach area. They can usually be detected by parting the wool. In areas of heavier infestation, there will be damaged fleece that becomes thin, ragged, and dirty from sheep biting and scratching. Also, sheep ked excrement will permanently stain the wool which can often be seen in heavy infestations.

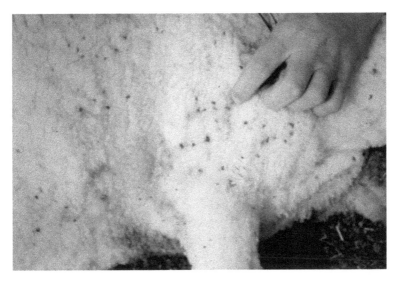

Figure 12.1
Sheep ked infestation. (Photo reproduced with permission by Jim Kalisch, University of Nebraska.)

Control

Spring shearing can reduce ked populations by up to 75%. This removes a significant number of pupae and adults. Also, flocks shorn prior to lambing will generally have lighter ked infestations than those shorn afterward.

Insecticidal treatment is most effective after shearing. All sheep in a flock must be treated, and all new animals should be isolated and treated before being mixed in with a flock. The application method used depends on the size of flock, facilities available, and labor. Methods available for insecticide application include sprays, dips, pour-ons, and dust application.

As a general rule, treat sheep in the spring after being sheared when the weather is warm and keds are fairly well exposed. However, if animals are heavily infested during the fall or winter months, consider treating at that time rather than allowing the keds to continue increasing, irritating and causing losses. If using sprays or dips during fall or winter, select warm, sunny days, treat animals in the morning, and keep them outside until they are dry. High-pressure sprays are desirable to penetrate the fleece and allow for total body coverage.

During cold weather, treating with pour-ons or dusts is recommended so as not to cold-stress animals. Power dusters can be effective for larger numbers of animals. A simple way to treat a few sheep for keds is to hand-dust them. Catch each animal, place it on its side, sift dust into the wool from the neck down the side to breech and on the belly, then rub the wool lightly by hand to work the dust down to the skin. Treat the opposite side the same. One treatment in spring or fall is usually sufficient to control infestations of sheep keds.

Figure 12.2
Sheep biting louse infestation.

LICE

Surveillance/Diagnosis

The most common indication of lice infestation on sheep and goats is animal rubbing, scratching, or biting at themselves. Make direct inspections on animals to confirm for the presence of lice. Lice can be found on most parts of the body, although the largest numbers are generally found in areas with longer fleece fibers (Figure 12.2). Chewing lice have broad brown heads and a pale brown body with dark bands. Chewing lice on sheep and goats are quite small. Sucking lice tend to be larger with narrow heads and much wider dark brown bodies. They sometimes appear bluish in color because of ingested blood.

Control

Control of lice on sheep and goats can be accomplished using the same procedures as for ked control.

SHEEP BOT FLY, OESTRUS OVIS (LINNAEUS)

Surveillance/Diagnosis

Animal behavior can be an indication of sheep bot fly activity. When the bee-like female flies are attempting larva depositing around the face of sheep, animals react by running or walking with their noses close to the ground or huddling in groups in circles with their heads down. They may also sneeze and stamp their feet and shake their heads. When larvae are in the nasal passages, nasal discharges are usually evident, often tinged with streaks of blood (Figure 4.10). In heavier infestations, animals show signs of difficulty in breathing. When larvae are migrating, sheep tend to sneeze profusely and often shake their heads.

Control

Control of the sheep bot fly is geared at larval infestations. Ivermectin, as an oral drench, is effective for controlling sheep bot fly infestations. The preferred treatment time is during late fall or early winter after adult fly activity has ceased. During this time, larvae in sheep are mostly first instars and are found primarily on the nasal mucous membranes.

WOOL MAGGOTS

Surveillance/Diagnosis

Certain species of calliphorid blow flies will lay their eggs in dirty wool or at wounds on sheep. Although the fly larvae do not feed on living tissue, they can cause significant irritation and skin and wool damage. Maggot-infested sheep become restless, stamp their feet, try to bite the irritated areas, and may leave the flock to hide in secluded places. Maggot infestation can be confirmed upon examination of wounds and in and around areas of soiled wool, especially around the hindquarters where feces and urine often stain wool. The wool overlying maggot infestation is discolored, moist, and foul smelling (Figure 12.3).

Control

Efforts should be taken to avoid blow fly strike (the egglaying activity of adult female blow flies). Keep animals as clean as possible. If the breech area becomes saturated with urine and feces during blow fly season, clip the wool from the crotch area and from the area above the tail down the back of the hind legs to the hocks. Prevent wounds by handling sheep gently and by providing safe chutes and corrals free of protruding objects.

Shearing early in the spring before fly season is a good practice. It removes soiled or fermenting wool, making sheep less attractive to flies, and permits shear cuts to heal before fly activity starts. Early lambing is also advisable for the protection of both ewes and lambs, because soiled wool of ewes from afterbirth and exposed umbilical cords of lambs may attract flies. When lambing occurs early, docking and castrating can be performed before fly

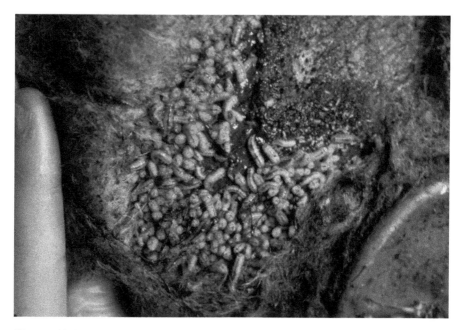

Figure 12.3
Wool maggot infestation.

season. Recently dead animals, fecal material, and rotting vegetation should be removed. Recently shorn sheep are seldom struck, and effective control of gastrointestinal parasites and foot rot, general animal health care, crutching, and trimming around the pizzle can further aid in the control of fly strike.

Insecticide treatments are useful for prevention as well as control of maggot or fly infestations. When sheep have accidental wounds or when necessary operations are performed during the fly season, timely application of a properly labeled insecticide directly onto the wound can be made. Preventative dips or sprays may be applied if animals become predisposed to fly strike during the warm months. Animals that become infested may be dipped or sprayed with insecticides or the maggot-infested wounds can be treated directly.

NUISANCE FLIES, MOSQUITOES, BLACK FLIES, BITING MIDGES, TABANIDS

In addition to wool maggot infestations, bot flies, and sheep keds, there are numerous other Diptera that are of concern to sheep. Blood-sucking Diptera, including stable flies, black flies, tabanids (horse flies, deer flies), biting midges, and mosquitoes can severely annoy sheep, cause blood loss, and may serve as vectors of disease agents. Also, house flies can be a source of nuisance.

Figure 12.4
Sheep reacting to black fly infestation.

Surveillance/Diagnosis

In and around sheep confinement facilities, stable fly and house fly activity can be monitored in several ways as discussed in Chapter 10. Light traps, baited fly traps, sticky fly ribbons, and spot (fly speck) cards as discussed for cattle can be employed. The selection of what to use and where to monitor for fly activity will vary according to the facility size and figuration.

Fly breeding spots in the sheep operation should also be monitored for. Designing a map or guide of the operation layout showing buildings, loafing areas, feeding stations, watering sources, walkways, and so forth, will be useful in the monitoring activity. Create a checklist of sites to monitor and keep this as part of a written fly management plan, along with adult fly activity records. This checklist can be used to assess the need for breeding site cleanup and maintenance and for fly control action steps needed.

With black flies, biting midges, mosquitoes, and tabanids, sheep behavior can be an indication of attack. Infested sheep tend to bunch up and refuse to feed or go to water. With black flies, especially, that attack the head and ears of sheep, animal bunching and pest avoidance behavior can be significant (Figure 12.4). Examination of sheep for fly bite damage can also be made.

Control

Sanitation/Cultural Practices   In controlling flies in and around barns and lots, sanitation is the first and most important step. No insecticide can be effective for flies around barns as long as breeding sites exist. This is especially true

259

in the case of the house fly and stable fly. All manure, spilled feed, wet straw, and decaying plant material should be removed weekly to break the breeding cycle of these flies, which is as short as 10 days from egg to adult fly. This can be done either by spreading manure and other waste material to dry or by placing it in pits or lagoons to become liquefied. If a liquid pit is used, do not allow accumulations of solid materials above the water line, either floating or sticking to the sides, because this would be conducive to fly production.

Populations of mosquitoes and biting midges can be controlled and sometimes eliminated by cultural water and habitat management practices. Manure-contaminated standing water near water tanks, animal pens, edges of lagoons, and low-lying areas should be eliminated or modified for proper water drainage and elimination of pest breeding habitat. Removal of emergent vegetation in drainage ditches and along edges of ponds and lagoons will help reduce mosquito breeding.

Mechanical Control: Screens and fly traps are two methods of mechanical fly control that can also be employed. Where possible, screen doors and windows should be used to prevent entry of flies. Many kinds of fly traps are available, most employing a black light with an electrically charged grill to kill the insects or baited with a fly-attractant material. Traps appear to be helpful in tight, enclosed areas where good sanitation practices are followed. However, in areas of heavy fly populations, traps are not effective in reducing fly numbers to satisfactory levels. Their best use is as a supplement to other fly control practices.

Biological Control: For confinement sheep operations, several insectaries have available parasitic wasps that can aid in biological fly control. Depending on the climate and geographic location, some species of these wasps are more effective than others in reducing fly numbers. The most effective species utilized are in the wasp family Pteromalidae and in the genera *Muscidifurax* and *Spalangia* (Figure 2.5). These are fly pupa parasites in which the female wasp oviposits one or more eggs inside a fly puparium. The larval wasp kills the developing fly and emerges from the dead fly puparium as an adult wasp. Natural populations of these parasites exist but usually do not develop in high enough numbers to provide noticeable fly control. Commercial release of these fly parasites can aid in the control of flies. It is advised to check with commercial suppliers of these parasites to customize a suitable program geared for a particular operation.

In using these fly parasites, for most effective fly control benefit the following conditions should be met:

- Proper sanitation and cultural practices must be carried out. Parasite releases complement these practices but cannot replace them.
- When insecticidal treatment is necessary for supplementing fly control, only use insecticides and application methods compatible with

parasite use. Nonresidual space sprays and fly baits, if applied properly, usually are safe to use.

Research studies have looked at other candidate biological control agents. Recently, a fungus, *Beauveria bassiana*, has shown promise as a fly control agent. This fungus works by attacking the cuticle of flies, leading to their death. Approved formulations of this fungus are currently being marketed for use in livestock and poultry operations.

Chemical Control: Insecticides can be used for fly control around sheep barns and operations. Residual insecticide sprays can be applied to fly resting surfaces. Several pyrethroid and organophosphate insecticides are available. Knockdown, nonresidual space sprays can also be applied. These should be used, as needed, when flies first appear in the spring. Several products are available as knockdown sprays containing such active ingredients as synergized natural pyrethrins or various pyrethroids.

Insecticide fly baits can be used. They are only effective against nonbiting flies, primarily house flies. Blood-feeding flies like stable flies are not attracted to baits. Most bait products available are in granular form and contain such active ingredients as methomyl, imidacloprid, and spinosad (Figure 2.18). Baits should only be applied in locations inaccessible to animals. They can be placed in bait stations, used as scatter baits in infested areas, or mixed with water to form a slurry (if allowed on the label) and applied with a brush on fly-infested surfaces. Baits are best used in buildings and enclosed areas and are less effective in outdoor areas.

Larvicides can be applied directly on manure and other fly breeding sources. However, their use should be reserved for treatment of fly breeding spots not eliminated by normal sanitation practices.

MITES

Several mites can infest sheep, including *Sarcoptes*, *Psorpotes*, and *Chorioptes* mites. Because production losses can be substantial with mite infestation, including weight gain losses and wool production losses, mite control in sheep is critical. Sheep mite infestations, especially sarcoptic mange and psoroptic scabies, are highly infectious. The severity of sarcoptic mange and psoroptic scabies requires infested animals to be quarantined and treated. With chorioptic mange, there is usually little damage to sheep, other than slight host irritation.

Surveillance/Diagnosis

An indication of mite infestation on sheep is visible lesion development and poor animal condition. To confirm infestation, skin scrapings need to be taken at infested sites on animals. With *Psoroptes ovis*, which causes sheep scabies, the lesions usually have the appearance of a dry scab surrounded by successive zones of moist crust and reddened skin. The mites are most active at the edge of the scab and are found on the skin surface. Using a scalpel and scraping along the surface of the edge of the scab can be done to confirm infestation. The scraped material should be examined using a microscope to confirm the presence of mites and to aid in species identification. *Chorioptes bovis* infestations on sheep can be sampled the same way because they also occur on the surface of the skin. With *Sarcoptes scabiei*, infested areas are characterized by hair loss and heavy, thick crusts or scabs. These mites feed on the moist layers under the scabs. Scrapings for sarcoptic mites need to be deep enough under the crusty scabs to expose the moist layers of skin. Scraping sheep for mites is best done by a veterinarian or other trained professional.

Control

In controlling mite infestations in sheep, treatments are more effective after sheep have been shorn. Dipping to soak animals sufficiently in acaricides (e.g., organophosphates) is an effective and approved method for treating for mite infestations. Also, systemic injectable treatments using ivermectin are effective. Repeated administration after 2 weeks may be necessary in heavy infestations. Psoroptic and sarcoptic scabies must be reported to state or USDA officials. In most states, infected flocks are quarantined until treated and pronounced free of infection. To prevent the occurrence of sheep scab, mange, or scabies, sheep and goat producers should avoid exposing a flock to infested animals. Animals suspected of being infested should be treated and isolated before mixing into a clean flock.

TICKS

Several ticks can be of concern in sheep and goat flocks in different regions of the country. Infestations of ixodid ticks (e.g., Rocky Mountain wood tick—*Dermacentor andersoni*, lone star tick—*Amblyomma americanum*, Gulf Coast tick—*A. maculatum*) and the argasid tick (e.g., the spinose ear tick—*Otobius megnini*) are often encountered. Tick infestation can lead to tick paralysis, tissue damage and secondary infections at tick bite sites, and resulting weight gain losses, wool production losses, and losses in milk production.

Surveillance/Diagnosis

The most feasible means of diagnosis of tick infestation is direct animal examination for attached ticks, especially during the times of year and in geographic regions with a history of tick activity.

Control

Controlling ixodid ticks on sheep and goats is best accomplished using approved acaricides (e.g., organophosphates, pyrethroids) with high-pressure spray treatments or by using dip-vats. Spinose ear ticks can be controlled by application of an acaricide dust or oil solution directly into the ear. In addition, ticks can be hand removed from animals, especially on animals affected by tick paralysis. If possible, removing animals from infested pastures can be done to prevent tick infestations.

MANAGEMENT OF POULTRY PESTS

TABLE OF CONTENTS

The U.S. poultry industry is the world's largest producer of poultry meat products and a major egg producer. According to the U.S. Department of Agriculture (USDA), there are over 340 million egglayers in the United States, and annual meat production of around 9 billion broilers, 270 million turkeys, and 27 million ducks. The total value of poultry production exceeds $32 billion annually.

Arthropod pests, including flies, beetles, and external parasites, cost poultry producers millions of dollars annually from reduced weight gains and meat quality, egg production losses, and structural damage. Costs are also associated with dealing with and minimizing nuisance complaints from fly and beetle activity originating from poultry production facilities. The shift from small farm flocks to larger commercial poultry operations has greatly increased pest concerns. The high-density, confined housing systems used in poultry production today create conditions that favor the development of manure-breeding flies and beetles associated with poultry litter accumulations. External parasites (especially mites and lice, and occasionally bed bugs, fleas, and ticks) are also of concern. In this chapter, discussion will focus on pest management strategies in controlling the various arthropods that affect poultry production.

FLIES

Several species of manure-breeding flies may be found associated with poultry production facilities including the house fly (*Musca domestica*), little (lesser) house fly (*Fannia canicularis* and *F. femoralis*), black garbage fly (*Hydrotaea aenescens*), black soldier fly (*Hermetia illucens*), blow flies (family Calliphoridae), small dung fly (family Spaeroceridae), and several other species of small gnats.

Accumulated poultry manure can be highly suitable for fly breeding, especially where general sanitation is poor and when there is excessive moisture. Suitable fly-breeding conditions can be present year round in enclosed high-rise egglayer houses with long-term manure accumulation and controlled indoor temperatures and in shallow pit houses in which manure is held for several months. In other types of poultry operations (e.g., breeder flocks, broiler/turkey grow-out houses), flies may also be of concern. In breeder houses with slatted floors, flies can find breeding spots, especially under feeders and waterers, and wherever manure accumulations under the slats have high enough moisture

Figure 13.1
House flies in a caged egglayer operation.

content. In broiler and grow-out houses for chickens and turkeys, little or no fly breeding usually occurs because the entire floor is covered with relatively dry litter. The litter is usually removed after three to six flocks and replaced with new wood shavings or other dry material. Some fly breeding may occur in wet litter around the waterers, but this is usually a minor problem.

The house fly is considered the major pest species associated with poultry manure, especially in caged-layer operations (Figure 13.1). House flies are the major cause of public health nuisance complaints in surrounding communities resulting in poor community relations and threats of litigation. The effective house fly dispersal range from their preferred breeding sources is from 0.5 to 2 miles, with nuisance populations highest closest to their breeding source. Of public health concern, house flies are capable of harboring more than 100 human and animal disease-causing organisms.

When *Fannia* sp. flies are present, high populations can develop on poultry farms. Little house flies are less tolerant of hot, midsummer temperatures than house flies, and they often emerge in large numbers in early spring, decline in midsummer, and peak again in late fall. Although these flies may invade nearby residential areas, they tend to be less annoying in that they do not readily settle on food or people. Adult males show a distinctive aimless hovering or circling flight behavior of long duration within the poultry house or at outside shaded areas. Females are less active and more often found near breeding sites.

Black garbage flies (*Hydrotaea aenescens* and *H. leucostoma*) can be found in large numbers in poultry facilities, breeding throughout the year. These

Figure 13.2
Black soldier fly larvae in manure. (Photo reproduced with permission by Christian Grantham.)

flies are generally considered to be beneficial, especially in enclosed egglayer houses. Black garbage fly larvae will actually kill house fly larvae and often dominate the manure habitat when present, especially in moist manure. Adult black garbage flies tend to stay on and around the manure surface in enclosed facilities. In poultry housing exposed to the outside, these flies are sometimes considered as nuisance pests.

The black soldier fly is common in poultry manure in some regions of the United States, especially in the southern, warmer states. The large, robust soldier fly larvae aggressively churn poultry manure (Figure 13.2). In doing this, they reduce manure volume and physically make this habitat less suitable for house fly and other muscoid fly larval development. In many situations, the black soldier fly becomes the dominant species present.

Several species of blow flies may occur in poultry facilities. They breed in decaying animal carcasses, dead birds, broken eggs, and wet garbage. Large numbers can be produced in a poultry operation if there are larger accumulations of broken eggs and if dead birds are not properly and frequently removed. Prompt removal of dead birds and rodents, prevention of the accumulation of broken eggs, and daily cleanup of processing areas are usually sufficient to prevent the buildup of these flies.

Small dung flies, along with several other small gnats, readily breed in poultry manure and other decaying materials. They can occur in large numbers in poultry operations but generally are not a nuisance on the farm or in nearby communities. Population levels are often higher in spring and late summer and fall.

Surveillance/Diagnosis

A standardized, quantitative method for monitoring fly populations should be a part of a poultry fly pest management program for use in making control

Figure 13.3
Fly fecal and regurgitation specks on egg.

decisions and to monitor control effectiveness. Visual observations of fly populations alone are subjective. Of sampling methods available, the use of spot/speck
cards, sticky fly ribbons, and/or baited jug traps are the most widely accepted.

Fly spot or speck cards consist of 3-inch by 5-inch white file cards placed
in a poultry house upstairs in high-rise or shallow-pit caged layer operations
and/or in the manure pit. They can be suspended from strings or fastened to
support posts, ceilings, or other areas where flies tend to settle (where there
are larger numbers of fly fecal and regurgitation spots) (Figure 13.3). Placement
is also best where there is little air movement and where workers or equipment will not disturb the cards. Several cards can be placed in a facility, with
date of placement and location noted on the card. Once placed, cards should
be left for a period of 7 days and replaced with new cards at the same place
each week. The number of "fly specks" on the exposed side (one side) of each
card should be counted and recorded in a record-keeping notebook or spread-
sheet. Generally, 100 or more fly specks per card indicate the need for fly
control measures. The use of spot cards is a simple, cost-effective, and widely
adapted method for assessing fly populations week after week. It also provides
documentation of fly activity over the course of time and the effectiveness of
control efforts that can be helpful in resolving conflicts with neighbors over
claims of increased fly activity.

Sticky fly ribbons/tapes are another means of monitoring fly activity in a
facility. One method to use them is to select locations to hang them up for

weekly intervals. However, used this way, they often tend to dry and get dirty over time and become less effective in capturing flies. A more suitable way to use them is to take a fresh tape, hold it out in front at waist level, and walk at a steady pace the length of the house down one walkway between cages and back another walkway (Figure 13.4). Flies caught on the tape can then be counted and recorded. One to two fly tapes should be used per house at least once a week. Generally, 100 or more flies caught per tape indicate the need for fly control measures.

Another method to monitor house fly activity is the use of baited-jug traps. These can be made from gallon plastic milk/food jugs (Figure 13.5). With each jug, cut four access holes 2 to 2.5 inches in diameter equidistant around the

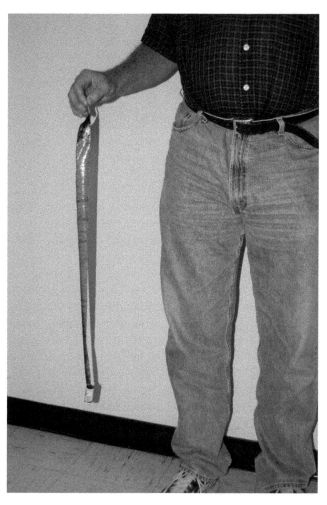

Figure 13.4
Sticky fly ribbon held at side. (Photo courtesy of Patrick Jones, Purdue University.)

Figure 13.5
Baited jug trap.

upper part of the jug. A wire or string is attached to the screwtop for hanging. About 1 oz of commercial fly bait containing fly attractant is placed on the inside bottom of the jug. Location of these traps is important for both effectiveness and accessibility. In high-rise houses and shallow-pit houses the traps can be hung from the ceiling or braces. In high-rise house pits, they should be hung 3 feet above the floor in areas not exposed to falling manure from above. Several traps can be placed through a facility. Traps should be examined weekly, the flies counted and removed, old bait discarded, and fresh bait added to the jug. An average of 250 flies trapped per week per jug indicates the need for fly control measures.

Fly Control

Successful fly control in poultry operations should be an integrated approach with emphasis on proper manure management. Four basic management strategies make up a successful integrated fly control program:

- Cultural/physical control
- Biological control
- Mechanical control
- Chemical control

Cultural/Physical Control   Management of poultry manure so that it is not conducive to fly breeding is the most effective means of control. Fresh poultry manure generally contains 60% to 80% moisture. Flies can readily breed in manure with a moisture content of 50% to 85%. Manure moisture below 50% is less suitable for fly breeding, and fly breeding usually does not occur at 30% moisture or less.

Dry manure management is practiced under two types of systems: frequent manure removal (at least weekly) and long-term, in-house storage of manure. Frequent manure removal systems to prevent fly breeding are based upon weekly (or more frequent) removal and field spreading it or transporting it to a holding area/composting site for drying/composting. This can be effective if done regularly and thoroughly, but it does require adequate and available agricultural land where manure can be spread or there are suitable facilities for holding manure or for composting. With belt manure removal systems, belts need to remove manure two to three times per week. Fresh manure on belts is attractive for fly egglaying activity, and fly maggots can develop if manure is not removed quickly enough. These developing flies may then continue developing in manure storage sites once the belt removes the manure. With in-house storage of manure, efforts should be made to reduce manure moisture below 50% (preferably to about 30% or less) and to maintain this level.

In either system, any practice that limits moisture in the droppings or aids rapid drying is helpful. A few practices to follow include the following:

- Prevent leaks in waterers. Inspect the pit daily to check for leaks, and repair them when found.
- When the water table is high or there is a danger of water running in from the outside, adjust the floor/grade relationship so that the floor of the house is higher than the surrounding ground and water runs away from the building.
- Provide abundant ventilation both in the manure pit for effective drying and in the house for bird comfort.
- Avoid rations that are laxative.
- Use absorbent litter where practical.
- Maintain proper insulation on water lines to prevent condensation.

Figure 13.6
Pit of caged egglayer house with dry manure coning.

With regard to ventilation, it reduces manure moisture and maintains desirable air temperatures, removes gases such as ammonia, and provides fresh air. Exhaust fans located in the manure pit walls provide ventilation for environmentally controlled high-rise houses. Also, supplemental drying fans (three-blade, 36-inch, direct-drive fans) installed between manure piles in the pits will greatly increase manure drying, especially that of the fresh dropped manure on the top of coned manure. These supplemental fans should be installed about every 50 feet, positioned in the same direction in each row, and reversed in position every other row to get maximum efficiency of air movement through the pit and manure surface. These fans should be kept free of obstructions and run during daylight hours when birds are most actively producing droppings.

In facilities designed for in-house storage of manure, accumulated droppings, if left undisturbed with adequate ventilation and free of additional moisture, will form a cone-shaped mound under the cages and allow for natural composting (Figure 13.6).

Undisturbed manure accumulations normally support large populations of beneficial parasites and predators of breeding flies. These parasite/predator populations primarily consist of predaceous beetles, mites, and parasitic wasps. The buildup of these natural fly enemies is usually slower than that of flies. Populations high enough to substantially benefit fly control can develop only if the manure is not disturbed for relatively long periods of time. To encourage parasites and predators,

- Maintain dry manure.
- Remove manure in cooler months when flies are less active.

- Stagger manure removal over a few weeks to preserve beneficial parasite and predator populations.
- Minimize the use of insecticides in the manure pit/storage area.

In-house composting of accumulating manure is a practice that can be established by producers to process manure and manage insect populations in high-rise layer houses. This process involves agitating the manure to incorporate oxygen and possibly a carbon source to aid in the composting process. This agitation results in increased temperature, an increased ammonia level, and decreased moisture content, all of which help reduce insect populations and make the manure more valuable as a fertilizer source that is easier to handle when removed. The agitation can be accomplished either by installing in-house chain- or tract-driven turning machines or by hydraulic-powered portable manure turners that can be used in multiple houses (Figure 13.7). Turning the manure pile twice a week is usually adequate, but the best way to determine need is to take the manure pile temperature with a compost thermometer. The temperature should be at least 120°F before turning is implemented. Because this composting activity produces ammonia that is released in higher concentrations during the turning process, protective respiratory devices should be used by workers. Also, composting eliminates beneficial parasite/ predator populations, so ceasing to compost after being initiated can cause large house fly outbreaks. Also, because most available composting equipment cannot continually turn high manure piles, the pit needs to be cleaned out

Figure 13.7
Portable manure turning machine.

when composted manure piles approach 2 feet in height for most available equipment.

When manure needs to be removed from a facility, it is best to do so in cooler months when flies are less active and when cooler outside conditions would minimize insect migration from field-applied manure. Also, fresh manure that accumulates within just 2 days after house cleanout is ideal for fly breeding, often resulting in a severe fly outbreak 2 to 3 weeks after a cleanout. When fly populations are low and less active during cooler weather, this fresh manure will be less prone to a quick establishment of fly breeding activity.

During spring and summer, when fly and beetle dispersal is a major concern, manure that must be removed can be treated with an insecticide approved for such use to kill insects in the manure at the time of removal. Manure can also be piled in a field before spreading and either treated again and/or tarped to kill developing insects. Heat that develops in the manure pile under secure tarps will kill insects that are present. After about 2 weeks under the tarp, the manure can be spread on fields without concern for pest dispersal.

Additional sanitation practices are also important in fly control. Remove dead birds daily and dispose of them properly. Minimize accumulation of spilled feed and broken eggs that attract flies and pest beetles. On the outside, keep grass and weeds adjacent to poultry houses mowed to eliminate resting areas for adult flies and to allow for adequate air movement around the buildings (Figure 13.8).

Biological Control   As indicated above, cultural/manure management practices encourage the survival and buildup of beneficial predators and parasites that can suppress house fly populations. Keeping manure dry also encourages the increase in other insects that compete for nutrients in the manure habitat.

Such beneficial organisms as predacious mite (e.g., *Macroceles* sp.) and small black hister beetles (*Carcinops pumilio*) (Figure 9.12) will readily feed on house fly eggs and first-instar house fly larvae. Another group of beneficial insects includes tiny parasitic wasps (primarily in the family Pteromalidae) (Figure 2.5). Female wasps oviposit their eggs in fly pupae. Inside the fly pupa, the developing larval wasp kills and consumes the fly before it emerges.

With proper dry manure management, predaceous mite and hister beetle populations often build up in higher numbers. Parasitic wasps (often called "parasitoids") usually occur naturally in lower numbers. Control using these parasitoids is sometimes based on mass releases of commercially reared parasitoids. Parasitoids are currently available from several commercial insectaries. For a release program to be successful, the producer needs to consider which parasitoid species are best suited for their particular operation and in what numbers to release them and when. Check with the suppliers of these parasitoids for recommendations.

Other insects, such as the darkling beetle (lesser mealworm, *Alphitobius diaperinus*) and dermestid beetles, often build up in high numbers under dry manure management (Figure 13.9). They can be beneficial in competing for the nutrients

Figure 13.8
Outside of a poultry house with cleared vegetation.

in the manure and prevent house fly buildup. However, they are responsible for damaging poultry structures (wood and insulation), harboring poultry disease organisms, and often being the cause of nuisance complaints when manure is transported and field applied when higher beetle populations are present in the manure. Control of these beetles is addressed later in this chapter.

A newer biological control agent has been developed commercially for fly control in poultry houses. A natural pathogenic fungus, *Beauveria bassiana*, has been formulated into a spray product (balEnce). This product is sprayed directly over accumulated manure. Adult house flies, as they emerge from their pupal cases, come in contact with spores of *Beauveria*. Spores released from the conidia develop hyphae that penetrate into the body cavity of the flies, resulting in death. *Beauveria* fungus has no detrimental effect on beneficial insects including histerid beetles and pteromalid parasitoid wasps.

Mechanical Control   Screens and fly traps are two methods of mechanical fly control, if used properly. Where possible, doors and windows should be screened to prevent entry of flies, especially in processing areas. Several kinds of fly traps are available. Some traps consist of a fly attractant in a liquid to attract flies, and others are electrical, employing a black light with either an electrically charged grid to kill the insects or sticky sheets to get attracted flies

Figure 13.9
*Large number of darkling beetles (*Alphitobius diaperinus*) in poultry waste.*

stuck. Traps do appear to be helpful in tight, enclosed areas where good sanitation practices are followed. However, in areas of heavy fly populations, traps are not effective in reducing fly numbers to satisfactory levels. They are best used as a supplement to other fly control procedures. Jug fly traps containing a liquid attractant can be strategically located around the outside perimeter of poultry buildings to help reduce adult flies near buildings. All fly traps need to be properly serviced and maintained to assure their optimal performance.

Chemical Control Insecticides should be considered as supplementary to sanitation and management measures aimed at preventing fly breeding. Producers should monitor fly populations on a regular basis to evaluate their fly management program and to decide when insecticide applications are needed. Chemical insecticides can play an important role in an integrated fly control program. However, improper timing and indiscriminate insecticide use can lead to increased fly populations. Also, selective application of insecticides can avoid killing beneficial fly predators and parasites. Insecticide applications may be directed to adult flies (adulticides) or fly larvae (larvicides). Methods of application include sprays (knockdown, residual), baits, and feed additives.

Knockdown, Nonresidual Space Sprays, Mists, and Fogs: These sprays are designed for quick knockdown and kill of flies with no residual action. They are usually the most effective and economical method to control potentially heavy populations of adult flies. Because they have very little residual activity, resistance to the insecticides recommended as space sprays is low, especially when using products containing synergized natural pyrethrins. Several pyrethroids are also available for use as quick knockdown sprays. There are many machines on the market designed to produce the small particle spray size desired for this type of application (e.g., backpack sprayers, cart-mounted sprayers) (Figure 2.15).

Space spray application should be made to the point of "filling" a room or area with the spray mist. Treatments should be made as frequently as needed to keep fly numbers down below identified nuisance levels. This method of fly control is best used in the cooler, early morning hours when flies are resting higher up in the house and ventilation fans can be safely turned off during the time of spraying without causing increased house temperatures. These insecticides should not be routinely applied in the pits of high-rise egglayer houses because they will also kill any beneficial insect populations present.

Residual Sprays: Treating building surfaces with residual sprays has been a common practice over the years. Dependence on this method has led to high levels of fly resistance to the available insecticides used as residual sprays (e.g., pyrethroids, organophosphates). Also, treated surfaces tend to quickly get covered over with dust, and this could reduce fly exposure on the treated surface. Residual sprays should be used sparingly and only as a last resort to control fly outbreaks that cannot be managed with other techniques.

Fly Baits: Baits are a viable part of an integrated fly control program to maintain low fly populations. They are a very effective supplement to sprays. Commercial dry baits in granular and extruded forms are readily available (Figure 2.18). They contain such active ingredients as methomyl, imidicloprid, and spinosad. Bait placement should be on walkways/aisles and other areas where flies congregate. Avoid application in the manure pit, because the available baits will kill beneficial parasite and predator populations. Baits must also be placed out of reach of birds and placed so they don't contaminate food and water sources. Some bait/insecticide products are also available on hanging strips. One such product, containing nithiazine, has been shown to be very effective and fast acting when properly placed in active fly areas. Also, some bait products are available as brush-on formulations or can be mixed with water to make a brushable slurry. Apply these treatments on surfaces where flies tend to congregate. Rotating the use of different bait products and active ingredients once or twice during a fly season will minimize the onset of fly resistance to any one active ingredient.

Larvicides: Direct application of chemical larvicides to the manure surface to kill fly maggots should be avoided, except for spot treatment or manure that is scheduled to be removed. This is especially so with products (e.g., pyrethroids, organophosphates) that will kill beneficial insects inhabiting the manure.

Cyromazine and pyridine spot treatments of small areas with higher numbers of maggots can be effective and yet have a minimal effect on the beneficial insect population and potential fly resistance development in the manure.

Feed-Through Larvicides: Cyromazine (Larvadex) is the only feed-through insecticide for breeding flies registered for caged layers. It is an insect growth inhibitor and kills fly larvae before development is completed. Its selective mode of action does not adversely affect natural fly predators. Larvadex premix is blended in to the egglayer ration at the rate of 1 pound of premix per ton of feed for fly control. It passes through the bird's digestive tract and is present in the manure essentially in its unaltered state. It has no adverse effect on feed palatability or consumption, or on eggs or meat.

Cyromazine will give best results when integrated into a well-managed fly control program. Use of this product too frequently can be expensive. Also, where it has been used extensively, high levels of fly resistance have been reported. It is best to use Larvadex after a complete manure cleanout. After cleanout, it can be fed to the birds continuously for 4 to 6 weeks. Its use after that should be avoided until the next cleanout. This will reduce the chance of development of fly resistance. If adult flies should become a problem during its use or after the time it is used, then proper adult fly control measures should be carried out.

BEETLES

Three species of beetles that commonly inhabit poultry manure and litter accumulations are the lesser mealworm or darkling beetle (*Alphitobius diaperinus*), the hide beetle (*Dermestes maculatus*), and the larder beetle (*D. lardarius*). Adults and larvae of these species, especially *A. diaperinus*, can become extremely abundant in poultry manure and litter, especially in drier accumulated poultry wastes (Figure 13.9).

On one hand, these beetles can be considered beneficial in that they compete in the same habitat as house flies and can help aerate and dry manure, making it unsuitable for house fly development. On the other hand, these beetles can cause extensive damage as mature larvae bore into structural materials seeking areas to pupate and complete their development. Mature lesser mealworms seek drier areas of the manure or litter and crack and crevice areas to pupate. They will bore into walls and can destroy house insulation, seeking areas to pupae (Figure 13.10). Over time, damaged insulation must be replaced because the loss insulating value results in greater heating costs as well as lowered feed conversion efficiency of the birds due to the lack of adequate temperature control in the houses. Mature hide beetle and larder beetle larvae often bore into wood posts, beams, and paneling to pupate (Figure 13.11). Over time, this could lead to weakened structures and building collapses. Even

Figure 13.10
Lesser mealworm damage to insulation.

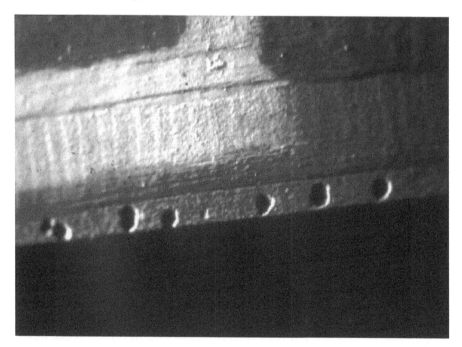

Figure 13.11
Hide beetle damage to wood.

slight partial collapses of wood support posts can cause movement of cages and equipment resulting in equipment failure.

These beetles are also known as potential vectors of several poultry disease pathogens (e.g., acute leukosis—Marek's disease, fowl pox, numerous pathogenic bacteria, and poultry tapeworms). In addition, large beetle populations, especially darkling beetles, may become a public nuisance at cleanout time because of adult beetle migration from fields where manure is spread into nearby residential areas.

Surveillance/Diagnosis

Monitoring beetle populations should be made in poultry houses in order to carry out control measures. Visual inspection for structural damage can be made to assess the extent of the beetle problem in a facility. In high-rise layer houses, pit inspections can be made to examine the manure directly for the presence of beetles. Beetle populations can often exceed 50 per square foot of manure surface in heavy infestations. In broiler and turkey growout houses, beetles are often under the litter and are not as easily noticed. They do tend to congregate in areas of higher temperature, suitable moisture (especially around waterers), and nutrients (mainly spilled feed). Consequently, greater numbers of beetles may be found under and around feeders and waterers in broiler and turkey growout houses.

In floor bird operations, and where desired in caged layer high-rise houses, beetle traps can be employed. Tube traps, consisting of rolled 12-inch-square corrugated cardboard inside a 12-inch-long section of 2-inch plastic pipe, can be used. They should be placed in areas where beetles tend to congregate, and where birds won't tamper with them. They should be checked weekly for beetle activity, replacing the corrugated cardboard if needed. If beetle population levels are established and significant damage to insulation and/or the structure has occurred, control efforts should be considered.

Beetle Control

In controlling beetles in infested poultry houses, applying dusts and/or sprays to manure and litter is fairly effective, but it can kill other beneficial insect populations. A thorough house cleaning, combined with chemical control when the birds are removed, will usually suppress beetle populations, at least for a short time. Migration within the poultry house may be reduced by applying insecticide sprays to pit walls and posts. Available insecticides include pyrethroids (cyfluthrin—Tempo, lambda-cyhalothrin—Demand), tetrachlorovinphos (Rabon), tetrachlorovinphos + dichlorovos (Ravap), carbaryl (Sevin), pyridine (Pyri-Shield, Archer), and boric acid.

During the time when manure is removed from a building, especially during warmer weather, efforts should be made to treat the manure to control developing flies and beetles. Treating the manure pit with one of the above insecticides a few days before it is removed will kill active stages of these insects. Once manure is removed from the building, if it is immediately field spread, treatment of the field may be necessary to kill surviving beetles. By stockpiling the manure first and treating the stockpile, this is a suitable alternative to get further beetle and fly kill. Thorough tarping of the stockpiled manure will also kill developing beetles and flies in the manure. A minimum of 2 weeks under the tarp will assure proper insect kill.

POULTRY LICE AND MITES

Several species of lice and mites make up the complex of external parasites of poultry. The physical damage caused by these pests may result in lowered egg product, reduced weight gain, and carcass downgrading. Also important is the nuisance to people handling eggs that are crawling with mites.

There are several species of chewing lice that may attack chickens, turkeys, ducks, and other domestic fowl (see Chapter 5). They are considered host specific and feed on poultry only, with their entire lifecycle spent on birds. With normal poultry management, lice have been seldom encountered except in small farm flocks or in floor-raised birds. However, occasional outbreaks in larger flocks have been reported.

The chicken mite, *Dermannysus gallinae,* feeds by sucking blood from birds at night. It hides in cracks and crevices in the poultry house during the day. They are more likely to be a problem in breeder houses where they have easy access to birds and adequate crack and crevice areas to hide and digest blood meals during the day.

The northern fowl mite, *Ornithonyssus sylviarum,* is the most important and common external parasite infesting poultry. It infests a wide variety of domestic fowl and wild birds. These mites spend their entire lifecycle on a host. They congregate near the vent, tail, and, occasionally, the back (Figure 7.3). Detailed information on these poultry mites is available in Chapter 7.

Lice and mites are introduced into poultry houses with the introduction of infested birds, or by wild birds, people, and infested equipment. Lice and northern fowl mites are normally transmitted from bird to bird by contact or simply by crawling to new hosts.

Surveillance/Diagnosis

In considering a pest management program for lice and mites, it is important to establish and maintain a monitoring program to assess when populations are detected and when economic thresholds warrant control. Although there is evidence to suggest that ectoparasite populations need to reach certain levels before economic loss will occur in poultry operations, there are several variables that could influence the need to control any one or more of these pests.

Housing design and condition, management system, bird density, ventilation, environmental stress, bird breed, bird age, type of bird production (e.g., egg-layer, broiler, breeder, etc.) may all contribute to the determination of the need to control. Another concern is the transport of ectoparasites from one house to another and from one farm to another that could introduce ectoparasites into a clean flock. At farms with multiple houses, the chance of spreading ectoparasites from one house to another becomes a real threat, especially if ectoparasite populations reach higher levels before treatment. Routine monitoring for the presence of lice and/or mites will assure detection when infestations occur and when they may be increasing.

Detection methods for lice, chicken mites, and northern fowl mites will vary because of differences in infestations and population levels that will develop with these parasites.

Poultry lice spend their entire lifecycle on birds. They are not as common on caged egglayers as they are on floor-reared birds. Data on the economic losses caused by lice are conflicting with some reports indicating little or no loss and others indicating lice can cause identifiable economic losses. The overall condition of the environment in which the birds are kept may influence how lice may or may not cause additional stress on birds. Thorough inspection of individual birds is the best means to detect lice infestations. Look for the presence of nymphs and adults and the buildup of eggs at the base of the feathers. The most common louse found in modern poultry production is the chicken body louse, *Menacanthus stamineus*. Eggs of this species can be readily seen in dense clusters at the base of the feathers when established infestations exist (Figure 13.12).

The chicken mite, unlike lice and the northern flow mite, is only found on birds when they feed, most commonly at night. They spend the rest of the time in cracks and crevices off of their host. Monitoring for the presence of chicken mites requires sampling of crack and crevice areas in the habitat where birds are kept. If birds show signs of stress of ectoparasite feeding (feather loss, skin damage, etc.) and no ectoparasites are found on the birds, then chicken mites could be the culprit. These mites need to be examined for in debris from roosts, slats, and nesting boxes and crack and crevice areas in the house, especially in breeder houses. Using a flashlight and a small brush, closely examine these concealed areas where the mites may be hiding. Although quite small, chicken mites can be seen upon close inspection. When they are found and birds appear to be affected, control should be considered.

Northern fowl mites are the target of most ectoparasite control efforts on poultry. Because northern fowl mites are continuous obligatory ectoparasites spending their entire lifecycle on birds, monitoring activity is focused on direct bird examination (Figure 7.3). Routine examination of a select number of birds in a flock on a weekly to monthly basis is vital to detect infestations and to monitor effectiveness of a control program.

The detection of an initial low northern fowl mite population that can be controlled effectively and economically is important in a mite-monitoring

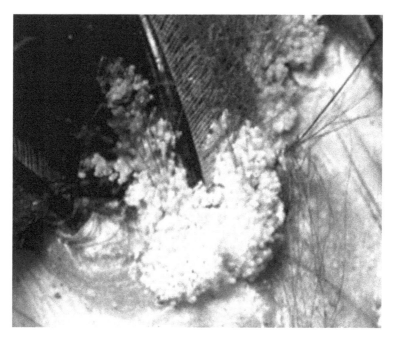

Figure 13.12
Chicken louse infestation.

program. With early detection, only part of the flock may need to be treated. At least 10 randomly selected birds from each cage row in caged-layer houses or from different sections of a floor operation should be monitored on a weekly to monthly schedule. The vent area should be examined using a bright flashlight or head lamp, and the feathers should be parted to reveal the mites. Cages with one or two birds often have more mites than those with more birds, and because of variation in susceptibility among birds, one bird may have mites while its cage mates have few or no mites.

The following index can be used for estimating infestation levels:

0 = no mites
1 = 1 to 50 mites (light infestation)
2 = 50 to 1000 mites—small clumps of mites on skin and feathers with beginning frass on some feathers (moderate infestation)
3 = 1000 to 25,000 mites—more frass accumulation on feathers and around vent (moderate to heavy infestation)
4 = 25000+ mites—numerous large clumps of mites on skin and feathers, with dense frass on at least 25% of feathers and skin pocketed with scabs (heavy infestation)

Generally, control efforts should be considered when index ratings of two or higher are detected.

Lice and Mite Control

Sanitation and cleanliness help prevent infestations of lice and mites. A poultry house should be clean and parasite-free before new birds are moved in. New birds should be checked and free of infestation before being brought in. Once a flock is in the facility, care should be taken to prevent contamination from workers and equipment. Mites and lice can be transferred from an infested house to an uninfested house by contaminated egg flats, bird crates, and other equipment. Chemical treatment (e.g., fumigation) of egg flats and cases is effective but inconvenient and costly. Wild birds and rodents can harbor and disseminate these parasites as well (especially mites). Wild birds should be excluded by screening, and rodents must be controlled in and around poultry houses. Detailed records of all flocks within a poultry operation should be maintained so that individual facilities with repeated mite infestations can be given special attention to prevent the spread of mites to other buildings and farms. Checking hatcheries and pullet operations for infestations and taking appropriate control steps in these facilities will often prevent infestations in egg production and grow-out facilities. In floor operations, before a previously infested house is restocked, it should be cleaned of all debris and feathers that may contain lice, mites, and their eggs. Also, a residual insecticide application should be made to the structure after being cleaned.

The decision to treat a flock is influenced by age of birds, time of year, and distribution of the infestation. It is not economical to treat older birds, because ectoparasite populations are unlikely to increase as compared to a younger flock. With lice and northern fowl mites, infestations are likely to increase in cooler months.

Chemical control of lice and northern fowl mites requires direct pesticide application to the bird, especially the vent region for northern fowl mites. Use sprays with sufficient pressure (100 to 125 psi) to penetrate the feathers. With caged birds, direct the spray upward from beneath the cages to adequately reach the vent area. Most pesticide "failures" result from inadequate treatment of the birds. This can be due to inappropriate formulations, poorly functioning application equipment, and the inability of the spray to reach the birds due to limited access in cage operations and the difficulties in herding the free birds in floor operations and breeder houses.

Depending on the chemical product used, wettable powders or emulsifiable concentrates mixed with water according to label directions are available for use. Emulsifiable concentrates mix better in water and are easier to use than wettable powders. Dust formulations may be applied with power dusters, and in the case of breeder flocks, the dusts may be applied directly in the nests. However, dusts do not usually give as satisfactory control as sprays. Dipping the birds and feather clipping can provide excellent control but are impractical and stressful to the birds. Insecticide-impregnated strips can be effective in caged layer houses, but the cost and difficulties in their installation make them impractical.

With chicken mites, because they spend most of their time off the host and the eggs are laid off the host, attention must be given to treating the structure and equipment. When an infestation is detected, not only should the birds be treated, but a thorough premise treatment should be made as well, using high-pressure sprays to penetrate crack and crevice areas where these mites may be found. Avoid contamination of bird food and water sources when making applications.

Chemicals available for lice and mite control include such organophosphates as malathion and tetrachlorovinphos, the carbamate carbaryl, and pyrethroids (e.g., permethrin). Northern fowl mite resistance has been reported with organophosphates and carbaryl from long-term use in some areas of the country.

BED BUGS

Bed bug, *Cimex lectularius*, infestations in poultry production facilities are sporadic, but they sometimes occur, especially in roost system operations and in breeder houses. Bed bugs, like chicken mites, will feed on birds primarily at night, spending their nonfeeding times in crack and crevice areas. Birds raised on litter or in cages are rarely infested because protective hiding places are lacking.

Surveillance/Diagnosis

Monitoring for the presence of bed bugs in poultry can be difficult but is done in a similar manner as for chicken mites. Birds need to be examined at night to find bed bugs feeding on them. Evidence on birds which indicates bed bug feeding activity is the presence of lesions on the breast and legs. Also, the presence of fecal spots on posts, nest boxes, and other surfaces will indicate their presence. Inspection for bed bugs on the premises should be done in the same manner as for chicken mites.

Control

Bed bug control starts with prevention of their being introduced and established in a poultry operation. The same cultural/sanitation procedures should be applied as for lice and mites. Cleaning and treatment of the house and equipment are critical for effective control.

If an infestation is found, a thorough chemical treatment using high-pressure sprays to penetrate hiding places is recommended. All houses on a farm should be treated at the same time to assure the most thorough control. The pesticides used for other ectoparasites are effective as sprays against bed bugs, especially permethrin. There is no documented resistance to these chemicals.

FOWL TICKS

Fowl ticks, *Argas* sp., are rarely encountered in modern poultry operations, and it is unlikely that a significant infestation would be encountered. If an infestation occurs, it would be more likely to occur in roost or breeder houses. With trends toward roost systems, they could become more of an issue, however. Control of fowl ticks, which are only found on birds when these ticks feed, would be accomplished by the same approaches used for the chicken mite and bed bug.

CHIGGERS

Chiggers that attack poultry are the same tiny larval stage of mites that attack people. Because chiggers are found on grass, weeds, and other low-growing vegetation, they are a problem to poultry that are raised outdoors on pasture, which is not a common practice today. To control chiggers on poultry on range or outdoor pens, carbaryl or malathion (sprays or dusts) can be used.

MANAGEMENT OF EQUINE PESTS

TABLE OF CONTENTS

The equine population in the United States is estimated at over 5 million head. This includes horses, ponies, mules, burros, and donkeys. The top ten states (Texas, California, Tennessee, Florida, Oklahoma, Pennsylvania, Ohio, Kentucky, Minnesota, New York) have a total of over 2 million head with an additional 15 states having inventories of more than 100,000 head per state. The total equine business, including animal value, land, and buildings, exceeds $20 billion. Annual expenditures for boarding, feed, tack, bedding, veterinary, and farrier costs often exceed $4000 per animal annually.

The U.S. horse and equine industry is very diverse. Horses are used for 4-H projects, recreational use, ranch and farm, breeding, racing, and equestrian sports. People of all ages obtain personal enjoyment and recreational benefits from equines. Because of the close relationship people have with horses and other equines, there is often more attention given to pest concerns as compared to production livestock.

Equine arthropod pests cost horse and other equine producers and owners more than $300 million annually from direct affects of arthropod loss and damage and in control costs. In addition, insect-borne diseases, including encephalitis, West Nile virus, and equine infectious anemia result in annual losses of more than $200 million to the equine industry. From 1999 through 2008, there have been over 25,000 equine cases of West Nile virus with over 15,000 cases alone in 2002.

Of concern to horses and other equines include numerous Diptera (house flies, stable flies, mosquitoes, horse and deer flies, black flies, biting midges, horse bots), lice, mites, ticks, and blister beetles. In this chapter, discussion will focus on pest management strategies in controlling these various arthropods that affect equines and their well-being.

FLIES

Most flies that are of concern with horses (and other equines) are also pests of other livestock. In pasture and outdoor areas, horses can be attacked and bothered by numerous kinds of flies. Face flies (*Musca autumnalis*) and horn flies (*Hematobia irritans*) can be of concern if horses are kept near cattle. Horse flies, deer flies, mosquitoes, biting midges, black flies, and eye gnats can also be of great annoyance, and, especially with mosquitoes, of grave concern with disease transmission. In stable and confinement areas, house flies (*Musca domestica*) and stable flies (*Stomoxys calcitrans*) are of primary concern. Horse bot flies (*Gastrophilus* sp.) are important as internal parasites.

Figure 14.1
Face flies on horse.

Face Fly, Musca autumnalis

Horses kept near cattle in temperate areas of the United States are readily attacked by face flies (Figure 14.1). These flies feed on the mucous tissues of the eyes that result in eye irritation. Face flies will also transmit eyeworms, *Thelazia* sp. in horses. Infestations of eyeworms are quite common in horses.

Diagnosis and Control   Observing behavior in horses will often indicate a face fly problem. Animals often become nervous and spend a lot of their time in shaded areas bunched together. Excessive irritation and secretions around the eyes are telltale signs of face fly feeding. *Thalazia* eyeworms can cause blockage of tear ducts resulting in redness and swelling of eye tissue.

Face fly control on horses consists primarily of insecticide wipes, mists, or sprays directed at the face (Figure 14.2). These treatments give horses temporary relief for a few hours. Horses perspire, and perspiration quickly breaks down insecticides. Also, most available insecticides contain pyrethrins and/ or pyrethroids that break down in direct sunlight. Because of this, daily treatments are needed for effective fly control and protection. Halter attachments and face masks are also available that can provide some protection of face fly attack. Overall reduction of face flies should be directed toward controlling flies on nearby cattle. See Chapter 10 for recommended control on cattle.

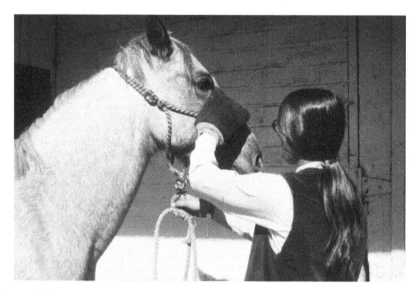

Figure 14.2
Wipe-on insecticide treatment on horse. (Photo courtesy of University of California.)

Horn Fly, Haematobia irritans

Blood-sucking horn flies will occasionally feed on horses kept near cattle outdoors. Horn fly populations on horses are easily controlled using a light insecticidal spray, dust treatment, or wipe-on application. These applications will protect horses from horn fly annoyance for short periods of time. A proper horn fly control program on cattle in the vicinity where horses are kept is an effective method of minimizing horn fly annoyance on horses.

House Fly, Musca domestica, *and Stable Fly,* Stomoxys calcitrans

House flies and stable flies are primarily of concern in and around horse confinement areas. House flies can be of annoyance to horses by their persistent buzzing and feeding around the eyes, muzzles, and open sores and wounds. Horses will become nervous and restless. House flies can also serve as intermediate hosts for the roundworm parasite of horses, *Habronema muscae*. In addition, if house fly breeding becomes excessive, this can be a source of fly nuisance to surrounding properties.

Blood-sucking stable flies can cause significant irritation to animals. They primarily feed on the legs of animals. Horses react by foot stomping and tail switching in an effort to dislodge the flies. As with house flies, stable flies can also transmit *Habronema* sp. nematodes to horses. The nematode is transmitted either through a feeding wound or internally if the horse swallows a fly harboring the nematodes.

House Fly and Stable Fly Control   Control of house flies and stable flies starts with sanitation. Bedding and manure should be removed at least weekly from horse stables, runways, paddocks, and arenas. It can be hauled to remote fields and spread thin to dry quickly. If stockpiled, it should be covered with black plastic to create enough heat to kill any developing flies and prevent moisture penetration.

Biological control of flies can be utilized on horse farms. Several insectaries have available parasitic wasps that can aid in fly control. Depending on the climate and geographic location, some species of these wasps are more effective than others in reducing fly numbers. The most effective species utilized are in the wasp family Pteromalidae and in the genera *Muscidifurax* and *Spalangia* (Figure 2.5). These are fly pupal parasites in which the female wasp oviposits one or more eggs inside a fly puparium. The larval wasp kills the developing fly and emerges from the dead fly puparium as an adult wasp. Natural populations of these parasites exist but usually do not develop in high enough numbers to provide noticeable fly control. Commercial release of these fly parasites can aid in the control of flies. It is advised to check with commercial suppliers of these parasites to customize a suitable program geared for a particular operation.

In using these fly parasites, for most effective fly control benefit, the following conditions should be met:

- Proper sanitation and cultural practices must be carried out. Parasite releases complement these practices but cannot replace them.
- When insecticidal treatment is necessary for supplementing fly control, only use insecticides and application methods compatible with parasite use. Nonresidual space sprays and fly baits, if applied properly, usually are safe to use.

Insecticides can be used as residual sprays or area/space sprays, baits, or as animal treatments. Residual sprays should be applied to fly resting surfaces (e.g., ceilings, walls, fences, sides of buildings, etc.). Organophosphate and pyrethroids available in residual spray formulations will usually be effective for up to 7 to 10 days if not washed off by rain or exposed to direct sunlight.

Space sprays with no residual activity (usually containing natural synergized pyrethrins or nonresidual pyrethroids) can be used as quick knockdown sprays for active adult fly control.

Ready-to-use granular or pelleted fly baits can also be used for house fly control (Figure 2.18). These need to be placed where flies are active but out of reach of horses. Some fly bait products are available as hanging strips. These also should be kept out of reach of horses if used.

On horses, insecticides can be applied directly to horses as sprays or wipes (Figure 14.2). It is best to first remove any excess dirt and dust on the animals. Apply the insecticide over areas of the animal to be protected, especially to the legs, belly, shoulder, neck, and face. It is recommended to treat the face with a wipe-on, keeping the insecticide away from the eyes and mucous

293

membranes. Check the label directions of any product used for retreatment intervals and any use restrictions. Daily treatment with some products may be necessary in heavy fly infestations. Some horse breeds have shown sensitivity to some insecticides. Check with a veterinarian if skin irritation occurs with insecticide application. There are many insecticide products on the market for fly control on horses. Most contain natural synergized pyrethrins or one of several pyrethroids. Check for a product that best suits the need for the fly control necessary.

One insecticide product Rabon® Oral Larvicide (tetrachlorovinphos) can be fed to horses in feeds or minerals to kill fly larvae that are breeding in feces. Use of this product is mostly effective against house flies that prefer manure as a breeding source. Stable flies often breed in nonmanure decomposing feed and vegetation and are not exposed to lethal amounts of this insecticide. Also, there can be a nonuniform dosage due to differences by individual animals in daily consumption resulting in some feces with sublethal levels of insecticide. Ivermectin, used as an internal parasite product, is also effective against house fly larvae breeding in the horses' feces as drug residues are expelled out in the feces. The residual effectiveness against fly larvae has been reported to last up to 30+ days after ivermectin treatment.

Horse and Deer Flies, Family Tabanidae

Horse and deer flies are strong fliers, vicious biters, and notorious pests of horses. They vary in size and color from the larger, robust horse flies to smaller deer fly species. Horses kept outdoors for any length of time are prone to attack by these flies. In addition to their painful bites, horse flies are also known vectors of the causative agent of equine infectious anemia.

Control of tabanids is difficult. Control of these flies in their aquatic breeding areas is impractical. Individual animal treatment with insecticides or repellents provides some protection from these flies if used on a daily basis.

Mosquitoes, Family Culicidae

Mosquitoes can be of significant importance to horses and other equine. Not only are mosquitoes of annoyance and can cause blood loss, but they are of major importance as vectors of the causative agents of several serious equine diseases. Mosquitoes can transmit equine infectious anemia, several strains of encephalitis that affect equines (eastern equine encephalitis, western equine encephalitis, Venezuelan equine encephalitis), and West Nile virus. With the threat of disease potential, mosquito control has become an important issue in the equine industry.

Mosquito control involves an integrated approach including source reduction, avoidance, and animal treatment. Source reduction focuses on the elimination and reduction of mosquito breeding sites. Mosquitoes require standing water for larval development. Any receptacle or area that holds standing water can be a source for mosquito breeding. Recommended examples to eliminate and reduce mosquito breeding include the following:

- Change water in water troughs at least once a week.
- Drain and fill stagnant pools, puddles, ditches, or swampy places around the property.
- Eliminate junk piles in outdoor areas.
- Destroy or dispose of tin cans, old tires, or any other unnecessary artificial water containers.
- Turn over wheelbarrows, pots, cans, and other receptacles that accumulate water. Add drain holes to those that tend to accumulate water.
- Keep rain gutters unclogged and flat roofs dry.
- Place tight covers over cisterns, cesspools, septic tanks, fire barrels, rain barrels, and tubs where water is stored.
- Fill all tree holes with sand or mortar, or drain them.
- Remove all tree stumps that may hold water.
- Keep margins of small ponds clear of vegetation.
- Ditch and clean stagnant streams to insure a continuous flow of water to eliminate border vegetation that produces habitat for mosquito larvae to develop.
- Drain or fill backwater pools and swamps where stagnant water accumulates.
- Stock small lakes and ponds with top-feeding minnows, if allowable.
- Improve wetlands and marshes to encourage development of mosquito predators (e.g., frogs, predatory insects, predatory fish).

Mosquito avoidance includes such practices as keeping horses inside stables at night and placing and maintaining screens on doors and windows to keep mosquitoes out. Also, the use of fans in stables can help to keep mosquitoes off of horses. And, because mosquitoes are attracted to lights, lighting in and around stables should be kept off during periods of mosquito activity.

There are numerous products available for use on horses to repel and/ or kill mosquitoes. Only products registered for use on equines should be used. The most common active ingredients in available products are synergized natural pyrethrins and various pyrethroids. Daily treatment with sprays or wipe-on applications is needed for the best protection. Thorough coverage on animals over their entire bodies affords the best protection against mosquitoes. Always read and follow label directions before using any product.

For disease prevention, there are vaccines available for encephalitis and West Nile virus. Typically, for West Nile virus, two doses of a vaccine are required. A booster is then required each year prior to the start of the mosquito season. A veterinarian should be consulted to determine individual animal needs for vaccination and/or treatment.

Black Flies, Family Simuliidae

Black flies can be severe pests of horses and other equines in some areas of North America. Their bites are quite painful and can cause pronounced itching and tissue irritation and swelling at the bite site. Although they may feed on all body areas of sparse hair covering, they are most troublesome to horses by feeding inside the ears. Severe scabbing and thickening of the tissue inside the ear often result (Figure 3.6). Horses become very irritated and head shy because of intense ear soreness.

Control of black flies off the host is not feasible. They breed in clean, swift-flowing streams that are not practical to treat. Area control of adult black flies is also not practical. Individual animal treatment with available insecticides can be effective for controlling black flies feeding in ears. An effective non-insecticidal treatment in the ears consists of smear applications of petroleum jelly inside the ears where black flies feed. This minimizes black fly feeding and lasts for about 3 days at a time.

Biting Midges (Family Ceratopogonidae) and Eye Gnats (Hippelates sp.)

There are several species of ceratopogonid biting midges that can feed on horses. Their painful bites can cause significant irritation and annoyance to horses. These small flies are often called "punkies" or "no-see-ums."

Eye gnats, primarily *Hippelates pusio*, can be commonly found around faces and open wounds of horses throughout the warmer months of the year. They can cause noticeable annoyance to animals.

For controlling biting midges and eye gnats, horses may be partially protected by frequent applications of repellents or insecticides, or confined to stables during periods of fly activity. Use of stable fans will provide some relief to horses to help keep these small flies off the animals.

Horse Bots, Gastrophilus sp.

Horse bots, as described in Chapter 4, are very common in horses (Figure 4.7). Effective control of horse bots requires breaking the lifecycle of the fly. For external control, a warm-water insecticide wash can be sprayed or sponged on areas where horse bot eggs are found (Figure 14.3). The warm-moist treatment causes the eggs to hatch and the larvae die from exposure to the insecticide. It is recommended to make these treatments weekly during peak oviposition periods in late summer. Mechanical removal of eggs before they hatch can be accomplished with grooming tools and hair clipping.

For internal control, oral treatments such as medicated feed, drenches, or pastes can be used. Ivermectin and some organophosphates are registered

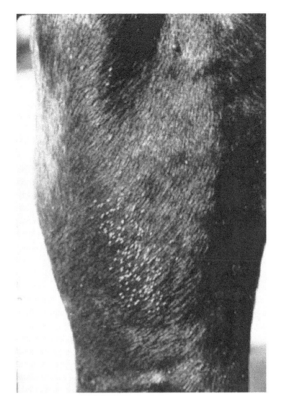

Figure 14.3
Horse bot eggs on horse.

for such use. These types of controls are most effective when the bots are in the stomach and after egglaying activity of adult females, generally from fall through mid-winter.

Cattle Grubs, Hypoderma *sp.*

Cattle grubs will occasionally infest horses, but they cannot complete their lifecycle in a horse. They sometimes can reach the back before they die. The presence of grubs in the back of horses, however, can be of concern to saddle horses, as the saddle will tend to irritate the infested area. Surgical removal of the encysted grub by a veterinarian is the best control method.

LICE

There are primarily two species of lice that are occasional pests of horses, the horse biting louse (*Bovicola equi*) and horse sucking louse (*Haematopinus*

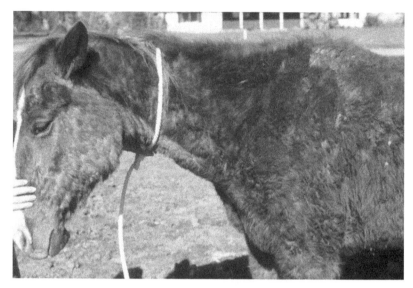

Figure 14.4
Horse infested with lice. (Photo courtesy of University of California.)

asini). They are both discussed in Chapter 5. Both species can cause irritation to horses and other equine. Infested animals will rub and scratch excessively, develop an unkempt coat, and may have patches of hair loss (Figure 14.4). Underfed and stressed animals, especially during the winter months, are more prone to severe louse infestations than animals maintained in good condition.

Lice Control

Proper grooming and feeding of horses is important to controlling lice infestations. Grooming and thoroughly inspecting an animal can often lead to the problem being spotted early before it becomes detrimental to the animal. Proper nutrition and animal care allow animals to better withstand the blood loss or irritation of a severe louse infestation. Horses should be routinely inspected for lice during the winter, especially if they are kept in close confinement with other animals.

It is best to treat animals individually by brisk, thorough brushing, wiping, or spraying the insecticide onto the animal, making sure the treatment penetrates the hair coat and covers the skin. Retreatment in 10 to 14 days may be necessary to kill young lice that hatch from eggs not affected by the first treatment. Most effective products available contain pyrethroid insecticides. Saddle blankets, brushes, rope halters, harnesses, and other tack should also be treated to prevent transfer of lice to clean hosts.

MITES

There are several mites that parasitize equines. The most common parasitic mite found is *Sarcoptes scabiei* var *equi*, which produces sarcoptic mange (Figure 14.5). Others include *Psoroptes ovis* (equi), producing psoroptic mange, *Chorioptes bovis*, producing chorioptic mange, and *Demodex equi*, the horse follicle mite. See Chapter 7 for details of the biology of these mites.

Sarcoptic mange in equine typically is first seen at the animal's withers and then spreads to the sides, back, shoulder, and neck. The entire lifecycle is spent on the animal with the female mite excavating tunnels in which she deposits her eggs. The burrowing causes scabbing at the skin surface. Animals will scratch, rub, and lick intensely at infested areas, causing unsightly lesions and hair loss. As spelled out in Chapter 11 for swine, diagnosis for sarcoptic mites is by examination of skin scrapings under a microscope.

In psoroptic mange (or scab), these mites do not burrow under the skin but feed externally. This causes an exudate that hardens to cover the feeding mites. They feed at the edge of the formed scabs. In equine, psoroptic mites usually first appear at the neck and withers, or at the base of the tail. If not controlled, the scabs enlarge and may cause debilitation and eventual death.

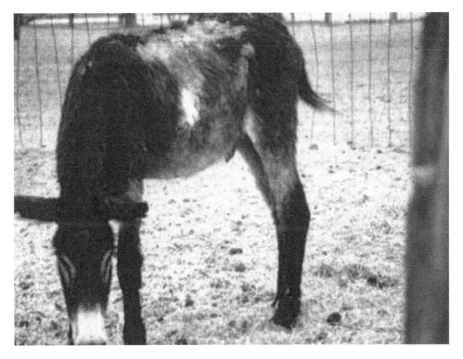

Figure 14.5
Sarcoptic mange in burro. (Photo courtesy of University of California.)

Chorioptic mange, often called tail mange, is most often observed around the tail. It spreads slowly, often extending to the legs (Figure 14.6). Mite infestations on the hocks produce severe irritation, loss of hair, and sores. Infested horses become restless. They will paw, lick, or bite at their lower legs in attempts to relieve the irritation.

The horse demodectic follicle mite is the most common mite found on horses in some areas. It lives in skin pores and is much smaller than the scab-producing mites. In clinical cases, demodectic mange in horses produces lumps or knots that form just under the skin resulting from pus that forms as a result of mite activity. Unless animals are carefully observed, infestations of this mite are normally undetected.

Horses are often pastured in areas infested with chiggers. These larval mites are not host specific and occasionally attack horses. In heavily infested areas, these mites can cause severe dermatitis in horses or other equine.

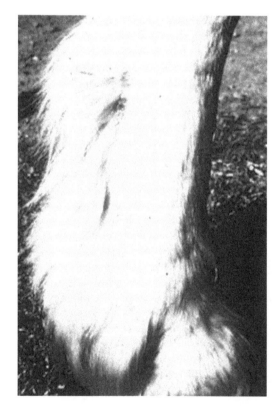

Figure 14.6
Chorioptic mange in horse. (Photo courtesy of University of California.)

Mite Control

Confirmation of sarcoptic and psoroptic mite infestations requires animals to be quarantined and proper control measures followed. These are federally reportable diseases. Chorioptic scabies is considered a reportable disease in some states. Reporting sarcoptic mange and psoroptic scabies is mandatory, and regulations specify the chemicals and methods of application to be used.

Early detection of mite infestation is important in limiting the spread to other animals and to prevent severe mange on animals. Spread of mites from animal to animal is by direct contact, or by the use of common grooming tools and tack. It is important to isolate infested animals and keep all equipment separated until the disease has been controlled. Verification of mange should be done by skin scrapings. Infested animals require thorough applications of approved acaricides. A high-pressure spray or thorough wetting with a brush/wash technique is necessary to saturate the skin to kill the mites. A second treatment in 7 to 10 days, depending on the product used, is recommended to control mites not killed by the first treatment. Chiggers on horses can be controlled with an application of an insecticide used for mange mites, ticks, or lice.

TICKS

Several species of ticks can be found infesting horses and other equine (Figure 14.7). Some tick species are widely distributed, whereas others are restricted to certain geographical ranges. Ticks are important because they are obligate ectoparasites, and they can transmit diseases to both humans and animals. Ticks vector protozoa, viruses, bacteria, rickettsia, spirochetes, and toxins, many of which can affect equine. Ticks of prevalence in the United States which often infest equine include the lone star tick (*Amblyomma americanum*), cayenne tick (*A. cajennense*), Gulf Coast tick (*A. maculatum*), winter tick (*Dermacentor albipictus*), American dog tick (*D. variabilis*), Rocky Mountain wood tick (*D. andersoni*), tropical horse tick (*D. nitens*), black-legged tick/deer tick (*Ixodes scapularis*), and spinose ear tick (*Otobius megnini*). The cattle fever tick (*Rhipicephalus annulatus*) and southern cattle tick (*R. microplus*) will also infest equine. With successful state and federal eradication efforts *R. annulatus* and *R. microplus* have been confined to occasional infestations in South Texas and southern California in buffer quarantine zones along the United States–Mexico border. Details on the veterinary importance, geographic distribution, and biology of all these ticks are presented in Chapter 8.

Tick Control

Ticks can be controlled on equine using approved acaricides. For effective control, acaricide sprays should be used with sufficient pressure to penetrate the

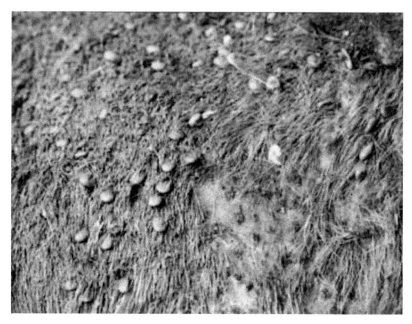

Figure 14.7
Tick-infested horse. (Photo courtesy of University of California.)

hair coat and wet the skin. Penetration into the ears is needed when ear ticks are a problem. Horse owners who frequently groom their animals can usually find and remove ticks on the horse before they become a serious problem.

BLISTER BEETLES (FAMILY MELOIDAE)

Blister beetles produce a toxic, irritating secretion, cantharidin. This chemical can cause severe dermal irritation and blistering to the skin and mucous membranes. Several species of blister beetles can be found on alfalfa. Adult beetles feed on alfalfa foliage and flowering parts and tend to congregate in small areas of the field. Because blister beetles congregate, a small number of alfalfa hay bales can be infested with many beetles. The amount of cantharidin varies from male to female beetles and among species of blister beetles. However, horses that eat as few as two to five blister beetles may be affected. Because of irritation and hemorrhaging in the digestive tract, horses can exhibit signs of colic, frequently void small amounts of blood-tinged urine, and at times have muscle tremors. Horses that eat more beetles may even die. Even the dried remains of whole beetles or crusted parts of beetles in alfalfa hay are toxic to horses.

For reducing the risk of feeding blister beetles to horses, the following steps should be taken:

- Use alfalfa hay from early season cuttings. First-cutting hay seldom has blister beetles present.
- Inspect alfalfa fields before cutting or baling.
- Cut alfalfa on a schedule that keeps alfalfa and weeds from producing the flowers that attract beetles. Cut before the advance of the bloom stage.
- Practice good weed management to keep other flowering plants to a minimum.
- Do not crimp or crush hay if beetles are present. Hay conditioning equipment will kill beetles as they pass through the rollers, contaminating the hay.
- Use sickle bar, circular, or rotary mowers that lay the alfalfa hay down but do not crush it. When the plants are disturbed this way, blister beetles fall to the ground and will leave to seek fresh food and moisture as the hay dries and cures.
- If possible, horse owners can grow their own alfalfa so that they can control all management practices and be sure the crop is beetle free.
- If not produced by the end user, do not use alfalfa hay from unknown sources. Buy from a local or known source and know what kind of management the hay has had.

CHAPTER 15

MANAGEMENT OF PET PESTS

TABLE OF CONTENTS

According to the American Pet Products Association (APPA) 2007 to 2008 statistics, 63%, or over 71 million, of households own a pet. Based on 2007 to 2008 statistics of the Humane Society of the United States, there are approximately 74.8 million owned dogs in the United States and 88.3 million owned cats. Approximately $9.8 billion is spent on pet supplies and over-the-counter (OTC) medication and $10.1 billion for veterinarian care annually. Annual routine veterinary care (not counting veterinarian visits) averages $219 per dog and $175 per cat according to 2007 to 2008 APPA statistics. Of this, a significant amount is spent on products to treat pets for arthropod pests (a majority of which is for flea and tick control). Of most concern to pets (primarily dogs and cats) include fleas, ticks, numerous Diptera (house flies, stable flies, myiasis-producing flies, mosquitoes, gnats), mites, and occasionally lice. In this chapter, discussion will focus on pest management strategies in controlling these various arthropods that affect pets and their well-being.

FLEAS (ORDER SIPHONAPTERA)

Although there are several kinds of fleas, only the cat flea (*Ctenocephalides felis*) is of primary concern as a pest of dogs and cats in the United States. The dog flea (*C. canis*) is not commonly found. Occasionally, wild animals (e.g., rodents) can bring in other flea species into a residence; however, these generally do not cause a concern for dogs or cats. See Chapter 6 for discussion of the veterinary importance and biology of fleas.

Surveillance/Diagnosis

Detecting flea activity starts by observing pet behavior. If a dog or cat scratches frequently, it is a sign of a flea infestation. Thorough examination of a pet can be made to confirm if fleas are present. Adult fleas are continuous obligatory parasites and once on a pet they prefer to stay. To assess the premise for adult flea activity, walk through suspect areas wearing exposed long white socks. Fleas present will readily jump onto the socks and can be counted. Walking through the entire residence will help to pinpoint areas of heavier flea activity. Focus should be made where pets sleep and spend a lot of their time. Check around pet bedding, cushioned furniture, on and around beds, thick carpeting, and protected areas where pets tend to sleep. Outdoors, "hot spots" of potential flea activity to check are dog houses, flower beds/gardens, under decks or porches, and other locations where pets spend time. Also check for activity of other cats and dogs, or other animals that may frequent the property.

Flea Control

Flea control involves an integrated approach including controlling adult fleas on pets and controlling developing flea populations off the host in flea breeding areas.

On Pets   On pets, control starts with confirmation of an infestation. A flea comb can be used on a regular basis to detect for and remove fleas from pets before infestations become well established. There are a multitude of over-the-counter and veterinary-supplied products on the market that can be applied as sprays, dusts, foams, shampoos, collars, feed additives, spot-on treatments, injectables, or pills. In the last few years, newer products on the market, especially veterinary-supplied products, have made it possible to kill virtually 100% of on-host flea populations and provide, in many cases, sustained residual efficacy for a month or longer with one treatment. Many of these products have become consumer-friendly in their ease of application, minimizing the need for environmental treatments.

Of over-the-counter products, insecticidal soaps, shampoos, powders/dusts, or spray-on liquids are still available. They vary in their effectiveness and are usually only effective in killing fleas at the time of treatment with limited residual effectiveness. Thorough body coverage is needed with these treatments.

Insecticide flea collars are also available. Depending on the product, they will provide varying degrees of flea control success. Some flea collars contain adulticides, and others contain insect growth regulators (IGRs) (e.g., methoprene, pyriproxyfen). These IGRs, when absorbed into the adult flea, will affect the flea's reproduction ability by causing eggs laid by the adult female not to hatch, thus breaking the flea lifecycle. Some flea collar products contain both an adulticide and an IGR.

The most effective on-host flea control products are usually veterinary supplied and formulated as spot-on treatments, tablets or feed additives for oral intake, or injectables. Topical spot-on treatments are applied at the animal's shoulder blades. The liquid treatment spreads over the animal's coat, providing whole-body treatment. They often provide effective flea control for 1 to 3 months with a single application. Active ingredients in spot-on formulations available that kill adult fleas include such insecticides as imidacloprid, permethrin (for dogs only), fipronil, metaflumizone, amitraz, selmectin, dinotefuran, and pyriproxyfen. Oral tablet products for adult flea control include such insecticides as nitenpyram and spinosad. Another active ingredient, lufenuron, is an insect development inhibitor. It is available in tablet form for dogs, as a liquid suspension to be mixed with a treat for cats, or as an injectable for cats. It works systemically being absorbed in the animal. Adult fleas pick it up in their blood meal. It effectively sterilizes the female flea and she lays eggs that do not hatch, thus breaking the lifecycle of the flea.

With any of these products, always check and follow label directions and the advice of a veterinarian as to proper dose to give, treatment intervals, which are recommended for dogs and cats, and if there are any restrictions for age of animal to treat or any animal breed restrictions.

Other products have been available for flea control on pests that do not work. Studies have shown that neither vitamin B1 (thiamine hydochloride) supplements nor brewer's yeast prevent fleas from feeding. Also, herbal-based pet collars and ultrasonic devices are not effective flea repellents.

Environmental Flea Control   With the advent of the effective topical and systemic products mentioned above for use on pests, existing flea populations can often be successfully eliminated without the need for premise control. However, treatment of premises may still be necessary where there are heavy flea infestations or there is severe pet or human allergy to fleas.

Flea control off the host starts with exclusion, source reduction, and proper sanitation. Efforts should be made to exclude and remove wildlife and feral animals to prevent the spread of fleas and prevent flea reinfestations onto a property. Remove pet food kept outdoors at night, secure fences and entryways, and close openings to crawl spaces to exclude animals from making dens under structures. Prevent wild animals, especially rodents, from getting into and nesting in houses or other buildings.

Inside homes, fleas can be controlled by good housekeeping practices. If pets have a history of flea problems, establish a pet sleeping area that can be cleaned easily and regularly. Fleas, as eggs, larvae, and pupae, can spend up to 90% of their development time off the pet. Before any insecticide is applied for flea control, it is advised to vacuum the premises thoroughly, especially pet resting areas, to remove flea eggs, larvae, and pupae. Vacuum carpets, rugs, upholstered furniture, and other areas where developing fleas may be present. Vacuuming on a weekly basis is recommended. This will also help pick up any shed hair skin particles from pets and other organic matter that flea larvae feed on. After vacuuming, seal and discard vacuum bag or canister contents soon after use to remove live fleas, eggs, larvae, and pupae picked up during vacuuming. Steam cleaning and/or shampooing will also reduce the number of developing fleas but usually will not eliminate a problem without additional control measures. Cleaning the carpet before applying insecticides lifts carpet fibers to allow for maximum penetration of the insecticide. Also, pet bedding, blankets, or rugs routinely occupied by pets should be cleaned on a regular basis.

Flea infestations in a premise usually become more evident when pets are removed. Although adult fleas in a premise prefer to feed on cats and dogs, when the pet is removed, fleas will readily feed on people. Dogs and cats can actually be used to attract fleas in a premise.

For premise treatments, the combined use of a residual insecticide and IGR will produce the best results. Common IGRs available include methoprene, hydroprene, and piriproxyfen. These can be used with residual sprays (e.g., pyrethroids, carbamates, organophosphates) that will quickly reduce adult flea populations. The IGRs will inhibit the development of the immature stages and disrupt the flea lifecycle. IGRs are available in premise spray and space spray formulations alone or in combination with residual adulticides. Some residual insecticides are also available as dusts. Insecticide sprays or dust should be applied as light, spot treatments to areas where flea activity is known to occur (e.g., in areas frequented by pets). These can be applied as surface treatments to floors, carpets, and upholstered furniture or can be used as crack and crevice treatments (Figure 15.1). Before spraying delicate fabrics, treat a small portion to be certain that the spray will not stain the fabric.

Figure 15.1
Spraying for fleas indoors.

Ready-to-use foggers can also be used. Some of the available products contain synergized pyrethrins for quick knockdown of adult fleas, whereas other foggers contain either an IGR, residual pyrethroid, or combination of both for longer-lasting control. When using total release foggers in a premise, several precautions should be followed. Remove all people and pets, including fish and birds, before treatment. Cover surfaces on which food is prepared. Place individual foggers in each room to be treated, placing them onto newspaper

Figure 15.2
Total release aerosol spray.

or a disposable surface (Figure 15.2). After treatment, the home must be thoroughly ventilated and chemical dried before people and pets return.

Most residual insecticide treatments will last for 2 to 3 weeks, with some of the IGRs lasting even longer. In heavily infested premises, retreatment may be necessary because flea pupae are protected from initial application of chemicals. Follow label directions as to when retreatment can be made.

For controlling flea infestations outdoors, it is first advised to thoroughly mow and rake a yard, and remove any organic debris from flower beds and gardens, under bushes, porches and crawl spaces, and in dog houses and kennels. This will increase exposure of fleas to insecticide treatment. Infested areas outdoors should be treated by applying a broadcast insecticide treatment (e.g., sprays, dusts) to infested areas (Figure 15.3). Application at 2- to 4-week intervals may be necessary for complete control. Both IGRs and residual insecticides (e.g., pyrethroids, organophosphates, carbamates) are available for outdoor use. Keep pets out of treated areas until dry.

TICKS

Pet animals, especially cats and dogs, are readily fed upon by ticks. How commonly ticks may be found on a pet varies depending on the region of the country, time of year, and habits of the pet. Cats and dogs that spend time outdoors frequently get ticks on them when ticks are active. The effect ticks may

Figure 15.3
Insecticide spray treatment for fleas outdoors.

have on pets will vary. When ticks attach and insert their mouthparts into the animal's skin, this can cause redness and irritation. The animal will chew and lick at the tick attachment site. Some animals may be so allergic to tick bites that they develop a "hot spot" causing severe inflammation and ulceration. Secondary infection may develop and continue to fester the tick bite site.

Some common ticks that may be found on pets outdoors, depending on geographic location, include the American dog tick (*Dermacentor variabilis*), lone star tick (*Amblyomma americanum*), and black-legged or deer

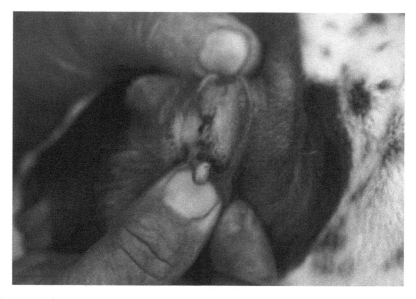

Figure 15.4
Brown dog tick infestation on dog.

tick (*Ixodes scapularis*). Other ticks that may occasionally infest cats and dogs include the Gulf Coast tick (*A. maculatum*), Pacific Coast tick (*D. occidentalis*), Rocky Mountain wood tick (*D. andersoni*), and spinose ear tick (*Otobius megnini*). The deer tick (*I. scapularis*) has become of special interest because it can transmit the causative agent of Lyme disease, *Borrelia burgdorferi*. Both dogs and cats are susceptible to Lyme disease. It can cause arthritis-like symptoms and swelling in the joints resulting in painful lameness. Depending on the tick species, other tick-borne diseases can be of concern, including Rocky Mountain spotted fever, tularemia, and tick paralysis. Indoors, the brown dog tick (*Rhipicephalus sanguineus*) can be of concern, primarily preferring dogs as a host (Figure 15.4). Discussion on the veterinary importance, distribution, description, and biology of these ticks is in Chapter 8.

Surveillance/Diagnosis

Regular, thorough examination of a pet can be made to confirm if ticks are attached or present (Figure 15.5). Adult ticks will stay attached for several days with the female engorging with blood. Often, the smaller adult male ticks may be found with the attached engorging females. Larvae and nymphs of some tick species, such as the lone star tick and deer tick, may also be found attached to pets. These immature stages are much smaller and may go unnoticed if the pet is not thoroughly examined. Off the host, a thorough examination around pet resting and bedding areas can be made to detect the presence

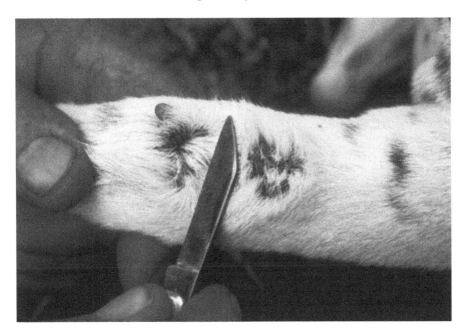

Figure 15.5
Examining dog for ticks.

of the brown dog tick. Close examination should be made of protected crack and crevice areas where nonfeeding stages of this tick often hide.

Tick Control

Ticks found on pets can be physically removed. If they are found attached, the following steps can be taken:

1. Use blunt forceps or tweezers.
2. Carefully grasp the tick as close to the skin as possible and pull upward with a steady, even pressure to remove the tick intact with its mouthparts.
3. Take care not to squeeze, crush, or puncture the tick.
4. Do not handle the tick with bare hands, because any infectious agents may be transferred onto mucous membranes or breaks in the skin.
5. After removing the tick, disinfect the bite site with a veterinarian-approved antiseptic and thoroughly wash hands with soap and water.
6. Consult a veterinarian if an infection develops.

For treating pets, some of the same products and formulations can be used as for fleas. Both cats and dogs can be sprayed with approved products containing pyrethrins. Topical spot-on treatments as applied for flea control containing fipronil or selamectin can be used on cats and dogs. Permethrin as a spot-on topical treatment can be used on dogs. *Do not use permethrin on cats.* Insecticide-impregnated pet collars containing propoxur are available for dogs and cats, and collars containing amitraz are available for dogs. These collars offer at least partial protection but may not provide total tick protection.

For controlling ticks off the pet in outdoor areas, efforts should first be made to keep overgrown and heavy vegetation cleared and cut in potential tick-infested areas. Eliminate unnecessary vegetation around yards or property, along wood edges, or along edges of trails and paths. Keeping grass and weeds cut short in tick-infested areas increases tick desiccation during hot weather, discourages habitation of alternative host animals, and lessens the amount of plant material that may need an insecticide application to kill ticks. Prevent unwanted wildlife, rodents, and stray dogs and cats from entering a property that could transport ticks. Also, remove clutter and debris to discourage habitation of rodents and other animals.

Residual insecticidal treatments can be applied on tick-infested areas, such as along property lines, roads, and animal trails where ticks congregate. Also, treatment can be made near ground level on grass and under shrubbery and trees, and at the edge of wooded areas. Available products contain such insecticides as tetrachlorovinphos, carbaryl, and several of the pyrethroids.

Indoor tick control, as well as in and around kennels, is directed primarily at the brown dog tick. Frequently clean or change pet bedding to prevent or remove ticks that may be found. Chemical treatments, when needed, can be applied in and around pet sleeping quarters, including around baseboards, window and door frames, cracks in walls, local areas of floors, floor coverings, along perimeter fences of kennels, and other areas where ticks may be found when not feeding on a host animal. Effective residual insecticide spray treatments available include various pyrethroids and carbaryl. Carbaryl is also available as a dust. Because tick eggs may hatch over several weeks, more than one treatment may be necessary to eliminate the problem. As a quick knockdown treatment to kill ticks present at the time of treatment, sprays containing pyrethrins can be used.

LICE

Lice infestations in dogs and cats are uncommon. When they do occur, they are highly host specific and more frequently encountered during the winter months. The cat biting louse (*Felicola subrostrata*) is the only species of lice that normally occurs on cats. Lice sometimes encountered on dogs include the dog biting louse (*Trichodectes canis*) and dog sucking louse (*Linognathus setosus*). Information on the veterinary importance, geographic distribution, description, and biology of these lice is discussed in Chapter 5.

Surveillance/Diagnosis

Cats and dogs infested with lice usually have dirty, matted, and odorous hair coats. Also, animals tend to bite, lick, and scratch profusely at infested areas. Upon thorough examination of the pet, lice can be seen at the skin and/or eggs can be seen attached to individual pet hairs. Other signs to look for are skin lesions, including scaly/scabby skin.

Lice Control

Upon positive identification, lice-infested cats and dogs can be effectively treated. Lice are continuous, obligatory parasites, spending their entire lifecycle on the host animal. It is only necessary to treat the cat or dog, not its surrounding environment. However, if there are other cats or dogs in regular contact with an infested animal, they should be treated as well.

Fipronil, as a spray or topical spot-on treatment, can be used to control lice on cats and dogs. Imidaclopeid and selamectrin administered according to label directions for flea control are also effective. Topical application of permethrin can be used on dogs. Do not use permethrin on cats. Treated animals should be placed in a clean cage or living space. Depending on product label directions, animals may be retreated in a week or so to ensure control of any nymphs hatching from eggs.

To prevent louse infestations, newly acquired cats or dogs should be thoroughly examined and treated if infested. Infested pets should be quarantined and treated before coming into contact with other pets. Keeping animals healthy and out of cold/wet weather, and maintaining them on an adequate plane of nutrition makes animals less susceptible to lice.

MITES

Dogs and cats may sometimes become infested with mites. Sarcoptic mange, demodectic mange, notoedric mange, and otodectic ear mites can become issues with pets. Also, in areas where chiggers can be a problem, pets can be affected as well as people.

Sarcoptes scabiei var. *canis* is somewhat common in dogs but less common in cats. It can cause intense itching, inflammation, and hair loss as these mites burrow under the skin. Crusty skin lesions will often develop. These mites prefer lightly haired regions and are more common on the elbows, hocks, ears, chest, and abdomen. In some extreme cases, the animal will lose large areas of hair and be covered with crusty scabs. Diagnosis of sarcoptic mange is by skin scrapings taken where lesions are found and by confirming the presence of the mites under microscopic examination. See Chapter 11 for procedures with swine.

Figure 15.6
Demodex *mange on dog.*

Demodex canis is a common mite on dogs found in hair follicles and seba-ceous glands. Most dogs maintain populations of this mite without showing clinical signs of mange. In some dogs, clinical mange can sometimes occur. Localized demodicosis occurs as isolated scaly bald patches, usually on the dog's face. It is more common in puppies and usually clears up without the need for treatment. In generalized demodicosis, sometimes called red mange, the entire dog can be affected with clinical symptoms of patchy hair, skin infections, and bald and scaly skin (Figure 15.6). Secondary bacterial infections make this a very itchy and smelly skin disease. Another form of demodectic mange in dogs is confined to the paws. Diagnosis of demodectic mange in dogs is by microscopic examination for the presence of mites in exudates from hair follicles and skin scrapings.

Notoedres cati affects cats very much like *Sarcoptes scabiei* affects dogs as it burrows under the skin. Symptoms in cats usually start with hair loss and scaly lesions on the ears and then spreading to the face, eyelids, neck, and shoulders. The mites will sometimes spread to the abdomen, legs and feet, and genital area, especially in younger animals. In infested areas, the skin becomes thick-ened, wrinkled, and covered with crusty scabs. Secondary infections often occur resulting from the intense irritation and scratching. Diagnosis is similar to that of sarcoptic mange in dogs by skin scrapings and microscopic examina-tion to confirm the presence of mites.

Otodectic ear mites (*Otodectes cyanotis*) typically will be found deep in the external ear canals of cats and dogs. Occasionally they will be found on other areas of the body including the head, neck, tail, and feet. They feed on tissue debris and secretions in the ear canal lining. Feeding irritation causes the ear canal to thicken, and reddish-brown to black debris from mite wastes and dead tissue resembling moist coffee grounds builds up in the ears. In established infestations, secondary infections may occur. Because ear mite infestations can be quite irritating, infested cats and dogs will scratch at their ears, shake their heads, or hold their heads to one side. Diagnosis for the presence of these mites can be made by examination with an otoscope or by recovering mites from the ear canal discharge using a cotton swab or aural scraping and observing mites with microscopic examination.

Mite Control

Mite control on cats and dogs starts with supportive nutrition. Pets kept on a high plane of nutrition often have a better resistance to mite attack. Carefully clip all long hair at infested sites and bathe the animal with a gentle cleansing shampoo (especially for sarcoptic and notoedric mange). Because sarcoptic and notoedric mange is very contagious, keep pets from coming in contact with stray or infected dogs or cats. Indoor pets are much less likely to contract infestation. Avoid boarding or grooming pets in locations that do not provide good sanitation and insist that grooming tools be disinfected between use. If a mite infestation is noticed, seek prompt treatment and isolate infested pets from other cats or dogs.

For demodectic mange in dogs, both oral and dermal treatments are available containing such compounds as ivermectin, amitraz, milbemycin, selamectin, and moxidectin. For sarcoptic mange control, most of these same treatments are available and effective.

In cats, lime sulfur baths and amitraz dips have been used to treat for notoedric mange. Ivermectin injections by veterinarians have become a common effective treatment.

For ear mite control in cats and dogs, removal of mite-infested debris from the ear can be accomplished using a few drops of mineral oil and gently massaging the base of the ear. This will loosen the exudate making it easier to remove. Although there are several over-the-counter products for ear mite control, the most effective products available can be obtained from veterinarians. Ivermectin, milbemycin, and thiabendazole are available as topical application formulations for administering directly into the ears. Selamectin and moxidectin are available as spot-on systemic treatments applied behind the pet's shoulders.

Follow veterinarian instructions and read and follow label directions before using any of these products for mite control. There may be restrictions for age or breed of pet. For example, ivermectin should not be used on collies,

Shetland sheep dogs, Australian shepherds, Old English sheep dogs, and some other herding breeds.

DIPTERA (FLIES, MOSQUITOES, GNATS)

Several Diptera are of concern to pets. Muscoid flies such as house flies, *Musca domestica*, can be a nuisance, and biting flies, especially stable flies, *Stomoxys calcitrans*, can cause significant irritation in their biting activity. In dogs, mosquitoes are of particular concern because of their role in transmitting dog heartworm. Other Diptera that bother pets, especially dogs, are *Hippelates* sp. eye gnats, ceratopogonid biting midges, and black flies.

Stable flies, sometimes called dog flies, can be of direct annoyance and irritation to dogs kept outdoors. They can viciously attack dogs, feeding around the ears causing raw, bleeding wounds. These blood-seeking flies seek this location because of the thin skin and hair on the ears and the inability of dogs to defend their ears. Dogs kept outdoors during the warm part of day will often experience hundreds of bites a day. This is especially so in kennels with outdoor runs and where there is adequate organic matter for these flies to breed in. Cats are much less likely to be attacked by stable flies.

Mosquitoes will also readily feed on pets that spend time outdoors. Certain species of mosquitoes, such as *Aedes* sp., will readily feed on mammalian hosts, including pets. These mosquitoes are also primary vectors of the nematode *Dirofilaria immitis*, the causative agent of dog heartworm. Dog heartworm has been reported in all 50 states of the United States, being more common in the eastern half of the country. Dogs and cats can get dog heartworm, but dogs appear more susceptible to infection. Mosquitoes become infected with heartworm microfilariae while feeding from an infected animal. These microfilariae mature in 10 to 14 days to the infective larval stage within the mosquito. When these mosquitoes feed on another dog or cat, the infective heartworm larvae enter through the bite site. They develop in the host animal for 6 to 7 months, maturing into adult worms. In dogs, these adult worms can live up to 7 years. The microfilariae produced cannot mature into heartworms without first passing through a mosquito. Heavily infected dogs will show clinical symptoms from coughing, fatigue, reduced appetite, and weight loss. Drugs, such as ivermectin and milbemycin, are effective preventive treatments for pets in oral, topical, and injectable formulations prescribed by veterinarians.

Other blood-feeding Diptera can also attack pets that spend time outdoors. Ceratopogonid biting midges and black flies (family Simuliidae) will readily feed on pets. Biting midges will feed on areas of the body where there is less hair, including the ears and underside of the animal. In late spring and early summer when black flies are prevalent, they will often be seen feeding on the ears of pets. These small biting flies can cause noticeable annoyance and irritation when they occur in larger numbers.

Hippelates sp. eye gnats can be of significant annoyance to pets, especially dogs, during the warmer months. These small flies that possess sponging mouthparts will readily feed on secretions around the eyes, mucous discharges, and moisture around the genitals of animals. They can become quite annoying to pets and people.

Blow flies (family Calliphoridae) sometimes can be of concern with pets causing secondary myiasis. If dogs or cats have open wounds or sores, insect or spider bites that become infected, or other skin trauma that may go unnoticed, adult calliphorid blow fly females may deposit eggs at these sites. Blow fly maggots will feed on the dead tissue at these wound sites. Blow fly larvae may also be found developing on thick-haired animals in areas that may accumulate urine or feces buildup on the hair coat. Although blow fly maggots will feed only on dead tissue, secondary infection may cause complications. One other form of myiasis that sometimes, but rarely, is seen in pets is that of the cuterebrid bot fly. Rodent bots in the genus *Cuterebra* have occasionally been found in dogs and cats. The bot fly larva will form a warble under the skin to develop (Figure 4.14). Once mature, it will exit the warble to pupate on the ground.

To prevent flies and other Diptera from bothering pets, keep animals indoors during peak insect activity. Various insecticides and repellent sprays and topicals are available for keeping flies off pets. If fly bites or blow fly maggot infestations are observed, gently cleanse the area with warm water and a mild antiseptic soap to soothe the bite or infested area. If the fly bites are severe, or extensive maggot activity is present, veterinary attention is needed. To minimize stable fly activity, keep kennels and other outdoor areas clean of potential fly breeding sources such as feces, uneaten pet food, dead vegetation, and garbage. Also, keep pets clean and well groomed, removing any urine-soaked or fecal-contaminated hair. If cuterebrid warbles are found on a pet, it is best to let the bot fly complete its development and leave the warble. The exit site can then be treated with an antiseptic to prevent secondary infection. If there are any complications with the pet, a veterinarian should be consulted. As with any insecticides applied on pets, always read and follow label directions for proper use and restrictions.

SELECTED REFERENCES FOR FURTHER READING

Allan, S. A., J. F. Day, and J. D. Edman. 1987. Visual ecology of biting flies. *Ann. Rev. Entomol.* 32:297–316.

Anderson, G. S., P. Belton, and N. Kleider. 1993. Hypersensitivity of horses in British Columbia to extracts of native and exotic species of *Culicoides* (Diptera: Ceratopogonidae). *J. Med. Entomol.* 30:657–663.

Arthur, D. R. 1962. *Ticks and disease.* Oxford: Pergamon.

Avancini, R. M. P., and M. T. 1990. Manure breeding insects (Diptera and Coleoptera) responsible for cestoidosis in caged layer hens. *J. Appl. Entomol.* 110:307–312.

Axtell, R. C. 1986. Fly control in confined livestock and poultry production. Technical Monograph. Ciba-Geigy, Greensboro, NC.

Axtell, R. C., and J. J. Arends. 1990. Ecology and management of arthropod pests of poultry. *Ann. Rev. Entomol.* 35:101–126.

Baker, A. S. 1999. *Mites and ticks of domestic animals. An identification guide and information source.* London: H. M. Stationery Office.

Baker, E. W., T. M. Evans, D. J. Gould, W. B. Hull, and H. L. Keegan. 1956. A manual of parasitic mites of medical or economic importance. Technical Publication of the National Pest Control Association.

Baker, E. W., and G. W. Wharton. 1952. *An introduction to acarology.* New York: Macmillan.

Baker, J. R., C. S. Apperson, S. M. Stringham, M. G. Waldvogel, and D. W. Watson, eds. 2000. *Insect and other pests of man and animals. Some important, common and potential pests in the Southeastern United States,* 2nd ed. Raleigh: North Carolina State University.

Balashov, Y. S. 1972. Bloodsucking ticks (Ixodoidea)—vectors of diseases of man and animals. *Misc. Publ. Entomol. Soc. Amer.* 8:161–376.

Barber, T., and M. J. Jochim. 1985. Bluetongue and related orbiviruses: Proceedings of an international symposium. Monterey, California, January 16–20, 1984. (Progress in Clinical Biological Research, 178). New York: A.R. Liss.

Baron, R. W., and D. D. Colwell. 1991. Mammalian immune responses to myiasis. *Parasitol. Today.* 7:353–355.

Beaty, B. J., and W. C. Marquardt, eds. 1996. *The biology of disease vectors.* Boulder: University Press of Colorado.

Bennett, G. W., J. M. Owens, and R. M. Corrigan. 2003. *Truman's scientific guide to pest management operations.* Cleveland: Advanstar Communications.

Benbrook, E. A. 1963. *Outline of parasites reported for domesticated animals in North America,* 6th ed. Ames: Iowa State University Press.

Bequaert, J. C. 1942. A monograph of the Melophaginae, or ked-flies, of sheep, goats, deer and antelopes (Diptera, Hippoboscidae). *Entomologica Americana (new series).* 22:65–220.

Bibikora, V. A. 1977. Contemporary views on the interrelationships of fleas and the pathogens of human and animal diseases. *Ann. Rev. Entomol.* 22:23–32.

Bishop, R. C., and H. L. Trembley. 1945. Distribution and hosts of certain North American ticks. *J. Parasitol.* 31:1–54.

Boreham, P. R. L., and R. B. Atwell, eds. 1988. *Dirofilariasis.* Boca Raton, FL: CRC Press.

Bram, R. 1978. Surveillance and collection of arthropods of veterinary importance. Washington DC: U.S. Department of Agriculture Handbook No. 518.

Braverman, Y., O. Marcusfeld, H. Adler, and B. Yakobson. 1991. Yellowjacket wasps can damage cow's teats by biting. *Med. Vet. Entomol.* 5:129–130.

Bruce, W. G. 1964. Then history and biology of the horn fly, *Haematobia irritans* (Linnaeus): with comments on control. North Carolina Agriculture Experiment Station Technical Bull. No. 157. North Carolina Agricultural Experiment Station and Entomology Research Division, Agricultural Research Service, U.S. Department of Agriculture, Raleigh.

Burger, J. F. 1995. Catalog of Tabanidae (Diptera) of North America north of Mexico. Contributions on Entomology International. 1:1–100.

Burgess, I. 1994. *Sarcoptes scabiei* and scabies. *Adv. Parasitol.* 33:235–293.

Busvine, J. R. 1969. *Lice,* 4th ed. London: British Museum of Natural History, Economic Ser. No. 2A.

Byrd, J. H., and J. L. Castner, eds. 2009. *Forensic entomology: the utility of arthropods in legal investigations,* 2nd ed. Boca Raton, FL: Taylor & Francis.

Carpenter, S. J., and W. J. LaCasse. 1955. *Mosquitoes of North America.* Berkeley: University of California Press.

Catanguie, M. A., J. B. Campbell, G. D. Thomas, and D. J. Boxler. 1997. Calculating economic injury levels for stable flies (Diptera: Muscidae) on feeder heifers. *J. Econ. Entomol.* 90:6–10.

Catts, E. P. 1982. Biology of New World bot flies: Cuterebridae. *Ann. Rev. Entomol.* 27:313–338.

Centers for Disease Control and Prevention (CDC). *CDC pictorial keys: arthropods, reptiles, birds and mammals of public health significance.* U.S. Department of Health and Human Services, Public Health Service, CDC. Washington, DC: U.S. Government Printing Office.

Cochrane, G. 1994. Effects of *Psoroptes ovis* on lamb carcasses. *Vet. Record.* 134:72.

Cole, F. R. 1969. *The flies of Western North America.* Berkley and Los Angeles: University of California.

Cole, N. A., and F. S. Guillet. 1987. Influence of *Psorptes ovis* on the energy metabolism of heifer calves. *Vet. Parasitol.* 23:285–295.

Collins, H. 1992. Control of imported fire ants: a review of current knowledge. U.S. Dept. Agriculture Tech. Bull. 1807.

Cooksey, L. M., and R. E. Wright. 1989. Population estimation of the horse fly, *Tabanus abactor* (Diptera: Tabanidae) in north central Oklahoma. *Environ. Entomol.* 16:211–217.

Cremers, H. J. W. M. 1985. The incidence of *Chorioptes bovis* (Acarina: Psoroptidae) on the feet of horses, sheep, and goats in the Netherlands. *Vet. Q.* 7:283–289.

Curran, C. H. 1965. *The families and genera of North American Diptera,* 2nd ed. Woodhaven, NY: Tripp.

DeFoliart, G. R., P. R. Grimstad, and D. M. Watts. 1987. Advances in mosquito-borne arbovirus/vector research. *Ann. Rev. Entomol.* 32:479–505.

DeVaney, J. A. 1978. A survey of poultry ectoparasite problems and their research in the United States. *Poultry Sci.* 47:1217–1220.

Drummond, R. O. 1987. Economic aspects of ectoparaites of cattle in North America. In W. H. D. Leaning and J. Guerrero, eds., *The economic impact of parasitism in cattle. Proc. of a Symposium, XXIII World Veterinary Congress, Montreal,* pp. 9–24. Veterinary Learning Systems, Lawrenceville, NJ.

Drummond, R. O., J. E. George, and S. E. Kunz. 1988. *Control of arthropod pests of livestock: a review of technology.* Boca Raton, FL: CRC Press.

Drummond, R. O., G. Lambert, H. E. Smalley, Jr., and C. E. Terrill. 1981. Estimated losses of livestock to pests. In D. Pimentel, ed., *CRC Handbook of pest management in agriculture,* Vol. 1, pp. 111–127. Boca Raton, FL: CRC Press.

Dryden, M. W. 1989. Biology of the cat flea, *Ctenocephalides felis felis. Comp Animal Prac.—Parasitol./Pathobiol.* 19:23–27.

Dryden, M. W. 1993. Biology of fleas of cats and dogs. *Comp. Cont. Educ. Pract. Vet.* 15:569–579.

Dryden, M. W. 1999. Highlights and horizons in flea control. *Comp. Cont. Educ. Pract. Vet.* 21:296–298, 361–365.

Dryden, M. W., and M. K. Rust. 1994. The cat flea: biology, ecology and control. *Vet. Parasitol.* 52:1–19.

Durden, L. A., and G. G. Musser. 1994a. The sucking lice (Insecta, Anoplura) of the world: a taxonomic checklist with records of mammalian hosts and geographical distributions. *Bull. Amer. Museum of Nat. Hist.* 218:1–90.

Durden, L. A., and G. G. Musser. 1994b. The mammalian hosts of the sucking lice (Insecta, Anoplura) of the world: a host-parasite checklist. *Bull Soc. Vector Ecol.* 19:130–168.

Elbel, R. E. 1991. Order Siphonaptera. In F. W. Stehr, editor. *Immature Insects.* Vol. 2:674–689, Kendell Hunt. Dubuque, IA.

Eldridge, B. F., and J. D. Edman, eds. 2000. *Medical entomology: a textbook on public health and veterinary problems caused by arthropods.* Dordrecht/Norwell: Kluwer Academic.

Emerson, K. C. 1956. Mallophaga (chewing lice) occurring on the domestic chicken. *J. Kansas Entomol. Soc.* 29:63–79.

Emerson, K. C. 1972. *Checklist of the Mallophaga of North American (north of Mexico).* Vols. 1–4. Dugway, UT: Desert Test Center.

Emerson, K. C., and R. D. Price. 1981. A host-parasite list of the Mallophaga of mammals. *Misc. Publ. Entomol. Soc. Amer.* 12:1–72.

Erzinclioglu, Y. Z. 1987. The larvae of some blowflies of medical and veterinary importance. *Med. and Vet. Entomol.* 1:121–125.

Evans, G. O. 1950. Studies on the bionomics of the sheep ked, *Melophagus ovinus* L., in West Wales. *Bull. of Entomol. Res.* 40:459–478.

Everett, A. L., I. H. Roberts, and J. Naghski. 1971. Reduction in leather value and yields of meat and wool from sheep infested with keds. *J. Amer. Leather Chemists' Assoc.* 66:118–130.

Everett, A. L., I. H. Roberts, H. J. Willard, S. A. Apodaca, E. H. Bitcover, and D. J. Haghski. 1969. The cause of cockle, a seasonal sheepskin defect, identified by infesting a test flock with keds (*Melophagus ovinus*). *J. Amer. Leather Chemists' Assoc.* 64:460–476.

Fallis, A. M. 1980. Arthropods as pests and vectors of disease. *Vet. Parasitol.* 6:47–73.

Ferris, G. F. 1951. The sucking lice. *Mem. Pacific Coastal Entomol. Soc.* 1:1–320.

Fincher, G. T. 1994. Predation on the horn fly by three exotic species of *Philonthus*. *J. Agric. Entomol.* 11:45–48.

Foil, L. D., and C. S. Foil. 1990. Arthropod pests of horses. *Comp. Cont. Educ. Pract. Vet.* 12:723–731.

Fox, I., and H. E. Ewing. 1943. The fleas of North America. U.S. Dept. of Agriculture Misc. Publ. No. 500. Washington, DC: U.S. Department of Agriculture.

Furman, D. P., and E. P. Catts. 1982. *Manual of medical entomology.* Cambridge: Cambridge University Press.

Furman, D. P., and E. C. Loomis. 1984. The ticks of California (Acari: Ixodida). *Bull. California Insect Survey* 25:1–239.

Geden, C. J., R. F. Stinner, and R. C. Axtell. 1988. Predation by predators of the house fly in poultry manure: effects of predator density, feeding history, interspecific interference and field conditions. *Environ. Entomol.* 17:320–329.

Georgi, J. R. 1990. *Parasitology for veterinarians,* 5th ed. Philadelphia: Saunders.

Gibbs, E. P. J., and E. C. Greiner. 1988. Bluetongue and epizootic hemorrhagic disease. In T. P. Monath, ed., *The arboviruses: epidemiology and ecology,* Vol. 2:39–70. Boca Raton, FL: CRC Press.

Gibbs, E. P. J., and E. C. Greiner. 1994. The epidemiology of blue tongue. *Comp. Immunol., Microbiol., and Infect. Dis.* 17:207–220.

Glasgow, J. P. 1963. *The distribution and abundance of tsetse.* New York: Macmillan.

Goddard, J. 2007. *Physician's guide to arthropods of medical importance,* 5th ed. Boca Raton, FL: Taylor & Francis.

Graham, O. H. 1985. Symposium on eradication of the screwworm from the United States and Mexico. Misc. Publication No. 62. Entomological Society of America, College Park, MD.

Graham, O. H., and J. L. Hourrigan. 1977. Eradication programs for the arthropod parasites of livestock. *J. Med. Entomol.* 13:629–658.

Graham, P. P. H., and K. L. Taylor. 1941. Studies on some ectoparasites of sheep and their control. 1. Observations on the bionomics of the sheep ked (*Melophagus ovinus*). Council of Scientific and Industrial Research, Pamphlet No. 108, pp. 8, 9–26. Melbourne, Australia.

Greenberg, B. 1971. *Flies and disease. Vol. 1. Ecology, classification and biotic associations.* Princeton, NJ: Princeton University Press.

Greenberg, B. 1973. *Flies and disease. Vol. 2. Biology and disease transmission.* Princeton, NJ: Princeton University Press.

Greiner, E. C. 1995. Entomological evaluation of insect hypersensitivity in horses. *Vet. Clin. N. Am., Equine Pract.* 11:29–41.

Greiner, E. C., V. A. Fadok, and E. B. Rabin. 1990 Equine *Culicoides* hypersensitivity in Florida: biting midges aspirated from horses. *Med. Vet. Entomol.* 4:375–381.

Grieve, R. B., J. B. Lok, and L. T. Glickman. 1983. Epidemiology of canine heartworm infection. *Epidemiol. Rev.* 5:220–246.

Guglick, M. A., C. G. Macallister, and R. Panciera. 1996. Equine cantharidiasis. *Compend. Con. Educ. Pract. Vet.* 18:77–83.

Guillot, F. F., and P. C. Stromberg. 1987. Reproductive success of *Psoroptes ovis* (Acari: Psoroptidae) on Hereford calves with a previous infestation of psoroptic mites. *J. Med. Entomol.* 24:416–419.

Hall, D. G. 1948. *The blow flies of North America*. Thomas Say Foundation, Entomological Society of America, College Park, MD.

Hall, M. 1995. Myiasis of humans and domestic animals. *Adv. in Parasitol.* 35: 257–334.

Hall, R. D. 1984. Relationship of the face fly (Diptera: Muscidae) to pinkeye in cattle: a review and synthesis of the relevant literature. *J. Med. Entomol.* 21:361–365.

Hanski, I., and Y. Cambefort, eds. 1991. *Dung beetle ecology*. Princeton, NJ: Princeton University Press.

Hardison, J. L. 1977. A case of *Eutrombicula alfreddugesi* (chiggers) in a cat. *Vet. Med. Small Anim. Clin.* 72:47.

Harwood, R. F., and M. T. James. 1979. *Entomology in human and animal health,* 7th ed. New York: Macmillan.

Heath, A. C. G., S. M. Cooper, D. J. W. Cole, and D. M. Bishop. 1994. Evidence for the role of the sheep biting louse, *Bovicola ovis*, in producing cockle, a sheep pelt defect. *Vet. Parasitol.* 59:53–58.

Helman, R. G., and W. C. Edwares. 1997. Clinical features of blister beetle poisoning in equids: 70 cases (1983–1996). *J. Am. Vet. Med. Assoc.* 211:1018–1021.

Hinkle, N. C., M. K. Rust, and D. A. Reierson. 1997. Biorational approaches to flea (Siphonaptera, Pulicidae) suppression—present and future. *J. Agric. Entomol.* 14:309–321.

Hoare, C. A. 1972. *The trypansomes of mammals*. Oxford: Blackwell Scientific.

Hogsette, J. A., J. F. Butler, W. V. Miller, and R. D. Hall. 1988. Annotated bibliography of the northern fowl mite, *Ornithonyssus sylviarum* (Canestrini and Fanzago), (Acari: Macronyssidae). Misc. Publication No. 76. Entomological Society of America, Lanham, MD.

Holland, G. P. 1964. Evolution, classification and host relationships of Siphonaptera. *Ann. Rev. Entomol.* 9:123–146.

Hollander, A. L., and R. E. Wright. 1980. Impact of tabanids on cattle: blood meal size and preferred feeding sites. *J. Econ. Entomol.* 73:431–433.

Hopkins, G. H. E., and T. Clay. 1952. *A checklist of the genera of species of Mallophaga*. London: British Museum of Natural History.

Hopkins, G. H. E., and T. Clay. 1952. *A check list of the genera and species of Mallophaga*. London: British Museum (Natural History).

James, M. T. 1947. The flies that cause myiasis in man. U.S. Dept. of Agriculture Misc. Publication No. 631. Washington, DC: U.S. Dept. of Agriculture.

Jones, C. J., and J. A. DiPietro. 1996. Biology and control of arthropod parasites of horses. *Comp. Cont. Educ. Pract. Vet.* 18:551–558.

Jones, C. J., and R. E. Williams, eds. 1989. Proceedings of a symposium: physiological interactions between hematophagous arthropods and their vertebrate hosts. Misc. Publication No. 71. Entomological Society of America, Lanham, MD.

Jones, R. H., A. J. Luedke, T. E. Walton, and H. E. Metcalf. 1981. Bluetongue in the United States: an entomological perspective toward control. *World Anim. Rev.* 38:2–8.

Jubb, K. V. F., P. C. Kennedy, and N. Palmer. 1985. *Pathology of domestic animals*. Vol. 3. San Diego: Academic Press.

Kahn, C. M., and S. Line, eds. 2005. *The Merck veterinary manual,* 9th ed. Rahway, NJ: Merial.

Keirans, J. E., and T. R. Litwak. 1989. Pictorial keys to the adults of hard ticks, family Ixodidae (Ixodida: Ixodoidea), east of the Mississippi river. *J. Med. Entomol.* 26:435–448.

Kettle, D. S. 1965. Biting ceratopogonids as vectors of human and animal diseases. *Acta Tropica.* 22:356–362.

Kettle, D. S. 1977. Biology and bionomics of blood sucking ceratopogonids. *Ann. Rev. Entomol.* 22:33–51.

Kettle, D. S. 1995. *Medical and veterinary entomology,* 2nd ed. Wallingford, UK: CAB International.

Kim, K. C., H. D. Pratt, and C. J. Stojanovich. 1986. *The sucking lice of North America. An illustrated manual for identification.* University Park: Pennsylvania State University Press.

Kirk, H. 1949. Demodectic mange. *Vet. Rec.* 61:394.

Kirkwood, A. C. 1980. Effect of *Psoroptes ovis* on the weight of sheep. *Vet. Rec.* 107:469–470.

Klein, K. K., C. S. Fleming, D. D. Colwell, and P. J. Scholl. 1990. Economic analysis of an integrated approach to cattle grub (*Hyopderma* spp.) control. *Can. J. Agricultural Econ.* 38:159–173.

Krafsur, E. S., and R. D. Moon. 1997. Bionomics of the face fly, *Musca autumnalis. Ann. Rev. Entomol.* 42:503–523.

Krafsur, E. S., C. J. Whitten, and J. E. Novy. 1987. Screwworm eradication in North and Central America. *Parasitol. Today* 3:131–137.

Krantz, G. W. 1978. *A manual of acarology.* Corvallis: Oregon State University.

Krinsky, W. L. 1976. Animal disease agents transmitted by horse flies and deer flies (Diptera: Tabanidae). *J. Med. Entomol.* 13:225–275.

Kwochka, K. W. 1987a. Fleas and related disease. *Vet. Clin. North Am., Small Anim. Prac.* 17:1235–1262.

Kwochka, K. W. 1987b. Mites and related disease. *Vet. Clin. North Am., Small Anim. Prac.* 17:1262–1284.

Lancaster, J. L., and M. V. Meisch. 1986. *Arthropods in livestock and poultry production.* New York: Halstead.

Lane, R. S., J. Peek, and P. J. Donaghey. 1984. Tick (Acari: Ixodidae) paralysis in dogs from northern California: acarological and clinical findings. *J. Med. Entomol.* 21:321–326.

Lane, R. S., W. Burdorfer, S. F. Hayes, and A. G. Barbour. 1985. Isolation of a spirochete from the soft tick, *Ornithodoros coriaseus*: a possible agent of epizootic bovine abortion. *Science.* 230:85–87.

Legg, D. E., R. Kumar, D. W. Watson, and J. E. Lloyd. 1991. Seasonal movement and spatial distribution of the sheep ked (Diptera: Hippoboscidae) on Wyoming lambs. *J. Econ. Entomol.* 84:1532–1539.

Legner, E. F. 1995. Biological control of Diptera of medical and veterinary importance. *J. Vector Ecol.* 20:59–120.

Lehane, M. 1991. *Biology of blood-sucking insects.* London: Chapman & Hall.

Lloyd, J. E., E. J. Olson, and R. E. Pfadt. 1978. Low-volume spraying of sheep to control the sheep ked. *J. Econ. Entomol.* 71:548–550.

Lloyd, J. E., R. E. Pfadt, and E. J. Olson. 1982. Sheep ked control with pour-on applications of organophosphorus insecticides. *J. Econ. Entomol.* 75:5–6.

Lofgren, C. S. 1986. The economic importance and control of imported fire ants in the United States. In S. B. Vinson, ed., *Economic impact and control of social insects*, Chapter 8, pp. 227–256. New York: Praeger.

Lofstedt, J. 1983. Dermatologic diseases of sheep. *Vet. Clin. North Am., Large Anim. Prac.* 5:427–448.

Loomis, E. C. 1986a. Ectoparasites of cattle. *Vet. Clin. North Am., Food Anim. Prac.* 2:299–321.

Loomis, E. C. 1986b. Insecticides and acaricides for cattle. *Vet. Clin. North Am., Food Anim. Prac.* 2:323–328.

Loomis, E. C. 1986c. Epidemiology and control of ectoparasites of small ruminants. *Vet. Clin. North Am., Food Anim. Pract.* 2:397–426.

Lowenstein, L. J., J. L. Carpenter, and B. M. O'Connor. 1979. Trombiculosis in a cat. *J. Amer. Vet. Med. Assoc.* 175:289–292.

Luedke, A. J., M. M. Jochim, and J. G. Bowne. 1965. Preliminary bluetongue transmission with the sheep ked *Melophagus ovinus* (L.). *Can. J. Comp. Med. Vet. Sci.* 29:229–231.

Marquart, W. C., R. S. Demaree, and R. B. Grieve. 1999. *Parasitology and vector biology*, 2nd ed. San Diego: Academic Press.

Marshall, A. G. 1981. *The ecology of ectoparasitic insects*. London: Academic Press.

Matthes, H. F. 1994. Investigations of pathogenesis of cattle demodicosis: sites of predilection, habitat, and dynamics of demodectic nodules. *Vet. Parasitol.* 53:283–291.

Matthyse, J. G. 1946. Cattle lice, their biology and control. Cornell University Agricultural Experiment Station, Ithaca. Bull. 832.

McAlpine, J. F., B. V. Peterson, G. E. Shewell, H. J. Teskey, J. R. Bockeroth, and D. M. Wood, eds. 1981. *Manual of Nearctic Diptera*. Vol. 1. Monograph No. 27, Research Branch, Agriculture Canada, Ottawa.

McAlpine, J. F., B. V. Peterson, G. E. Shewell, H. J. Teskey, J. R. Bockeroth, and D. M. Wood, eds. 1987. *Manual of Nearctic Diptera*. Vol. 2. Monograph No. 28, Research Branch, Agriculture Canada, Ottawa.

McAlpine, J. F., B. V. Peterson, G. E. Shewell, H. J. Teskey, J. R. Bockeroth, and D. M. Wood, eds. 1989. *Manual of Nearctic Diptera*. Vol. 3. Monograph No. 32, Research Branch, Agriculture Canada, Ottawa.

McDade, J. E. 1990. Ehrlichiosis—a disease of animals and humans. *J. Infect. Dis.* 161:609–617.

Meleney, W. P., and K. C. Kim. 1974. A comparative study of cattle-infesting *Haematopinus* with redescription of *H. quadrpertusus* Fahrenholz, 1919 (Anoplura: Haematopinidae). *J. Parasitol.* 60:507–522.

Messinger, L. M. 1995. Therapy for feline dermatoses. *Vet. Clin. North Am., Small Anim. Pract.* 25:981–1005.

Metcalf, R. L., and R. A. Metcalf. 1993. *Destructive and useful insects: their habits and control*. New York: McGraw-Hill.

Meyer, N. V. 1996. *History of the Mexico–United States screwworm eradication program*. New York: Vantage Press.

Miller, W. H. 1984. Diseases of domestic animals. In W. B. Nutting, ed., *Mammalian diseases and arachnids*, Vol. 2, Chapter 6, pp. 115–126. Boca Raton, FL: CRC Press.

Morgan, C. E., and G. D. Thomas. 1974. Annotated bibliography of the horn fly, *Haematobia irritans* (L.), including references on the buffalo fly, *H. exigua* (de Meigere), and other species belonging to the *Haematobia*. Agricultural Research Service, Misc. Publication No. 1278. pp. 1–134. U.S. Dept. of Agriculture, Washington, DC.

Morgan, C. E., and G. D. Thomas. 1977. Supplement I: annotated bibliography of the horn fly, *Haematobia irritans* (L.), including references on the buffalo fly, *H. exigua* (de Meigere), and other species belonging to the *Haematobia*. Agricultural Research Service, Misc. Publication No. 1278. pp. 1–38. U.S. Dept. of Agriculture, Washington, DC.

Morgan, C. E., G. D. Thomas, and R. D. Hall. 1983a. Annotated bibliography of the stable fly, *Stomoxys calcitrans* (L.), including references on other species belonging to the Genus *Stomoxys*. Missouri Agriculture Experiment Station Bull. No. 1049. University of Missouri, Columbia.

Morgan, C. E., G. D. Thomas, and R. D. Hall. 1983b. Annotated bibliography of the face fly, *Musca autunmalis* (Diptera: Muscidae). *J. Med. Entomol.* Supplement 4:1–25.

Moriello, K., and I. S. Mason. 1995. *Handbook of small animal dermatology.* London: Permagon Press.

Mullen, G., and L. Durden, eds. 2002. *Medical and veterinary entomology.* New York: Academic Press.

Muller, G. H., R. W. Kirk, and D. W. Scott. 1989. *Small animal dermatology.* Philadelphia: W. B. Saunders.

Mundell, A. C. 1990. New therapeutic advances in veterinary dermatology. *Vet. Clin. North Am., Small Anim. Pract.* 20:1541–1556.

Nelson, W. A. 1958. Transfer of sheep keds, *Melophagus ovinus* (L.) from ewes to lambs. *Nature.* 181:56.

Nelson, W. A., and A. R. Bainborough. 1963. Development in sheep of resistance to the ked *Melophagus ovinus* (L.). III. Histopathology of sheep skin as a clue to the nature of resistance. *Experimental Parasitol.* 13:118–127.

Nelson, W. A., and S. B. Slen. 1968. Weight gains and wool growth in sheep infested with the sheep ked *Melophagus ovinus. Exp. Parasitol.* 22:223–226.

Norris, K. R. 1965. The bionomics of blow flies. *Ann. Rev. Entomol.* 10:47–68.

Nutting, W. B. 1976. Hair follicle mites (*Demodex* spp.) of medical and veterinary concern. *Cornell Veterinarian.* 66:214–231.

Oliver, J. H., Jr. 1989. Biology and systematics of ticks (Acari: Ixodidae). *Ann Rev. Ecol. Systematics.* 20:397–430.

Oldroyd. H. 1964. *The natural history of flies.* London: Weidenfeld and Nicolson.

Olsen, O. 1974. *Animal parasites, their life cycles and ecology.* New York: Dover.

Peterson, J. J., and G. L. Greene, eds. 1989. Current status of stable fly (Diptera: Muscidae) research. Misc. Publication No. 74. Entomological Society of America, Lanham, MD.

Pfadt, R. E., L. H. Paules, and G. R. DeFoliart. 1953. Effect of the sheep ked on weight gains of feeder lambs. *J. Econ. Entomol.* 46:95–99.

Pfadt, R. E., J. E. Lloyd, and E., W. Spackman. 1973. Control of insects and related pests of sheep. Agriculture Experiment Station Bull. 514R. University of Wyoming, Laramie.

Pickens, L. G., and R. W. Miller. 1980. Biology and control of the face fly, *Musca autumnalis* (Diptera: Muscidae). *J. Med. Entomol.* 17:195–210.

Pittaway, A. R. 1992. *Arthropods of medical and veterinary importance: a checklist of preferred names and allied terms.* Tucson: University of Arizona Press.

Price, M. A., and O. H. Graham. 1997. Chewing and sucking lice as parasites of mammals and birds. Technical Bulletin No. 1849, U.S. Department of Agriculture.

Rafferty, D. E., and J. S. Gray. 1987. The feeding behaviour of *Psoroptes* spp. mites on rabbits and sheep. *J. Parasitol.* 73:901–906.

Ristic, M. 1988. *Babesiosis of domestic animals and man.* Boca Raton, FL: CRC Press.

Rothschild, M. 1975. Recent advances in our knowledge of the order Siphonaptera. *Ann. Rev. Entomol.* 20:241–259.

Rueda, L. M., and R. C. Axtell. 1985. Guide to common species of pupal parasites (Hymenoptera: Pteromalidae) of the house fly and other muscoid flies associated with poultry and livestock manure. Tech. Bull. No. 278. North Carolina Agriculture Research Service, Raleigh, NC.

Rust, M. K., and M. W. Dryden. 1997. The biology, ecology, and management of the cat flea. *Ann. Rev. Entomol.* 42:451–473.

Sargison, N. 1995. Differential diagnosis and treatment of sheep scab. *In Practice.* 17:3–10.

Schmidt, C. H., and J. A. Fluno. 1973. Brief history of medical and veterinary entomology in the USDA. *J. Wash. Acad. Sci.* 63:54–60.

Scholl, P. J. 1993. Biology and control of cattle grubs. *Ann Rev. Entomol.* 39:53–70.

Scholl, P. J., D. D. Colwell, J. Weintrub, and S. E. Kunz. 1986. Area-wide systemic insecticide treatment for control of cattle grubs, *Hypoderma* spp. (Diptera: Oestridae): two approaches. *J. Econ. Entomol.* 79:1558–1563.

Schmitz, D. G. 1989. Cantharidin toxicosis in horses. *J. Vet. Intern. Med.* 3:208–215.

Schwinghammer, K. A., F. W. Knapp, J. A. Boling, and K. K. Schillo. 1986a. Physiological and nutritional response of beef steers to infestations of the horn fly (Diptera: Muscidae). *J. Econ. Entomol.* 79:1010–1015.

Schwinghammer, K. A., F. W. Knapp, J. A. Boling, and K. K. Schillo. 1986b. Physiological and nutritional response of beef steers to infestations of the stable fly (Diptera: Muscidae). *J. Econ. Entomol.* 79:1294–1298.

Scott, D. W. 1983. *Large animal dermatology.* Philadelphia: W. B. Saunders.

Scott, D. W., R. D. Schultz, and E. B. Baker. 1976. Further studies on the therapeutic and immunological aspects of generalized demodectic mange in the dog. *J. Amer. Anim. Hospital Assoc.* 12:202–213.

Sheppard, D. C. 1983. House fly and lesser house fly control utilizing the black soldier fly in manure management systems for caged laying hens. *Environ. Entomol.* 12:1439–1442.

Sheppard, D. C., G. L. Newton, S. A. Thompson, and S. Savage. 1994. A value added manure management system using the black soldier fly. *Bioresource Technol.* 50:275–279.

Skidmore, P. 1985. *The biology of the Muscidae of the world.* Dordrecht: W. Junk.

Snow, J. W., A. J. Siebenaler, and F. G. Newell. 1981. Annotated bibliography of the screwworm, *Cochliomyia hominovorax* (Coquerel). U.S. Dept. of Agriculture, ARM–S-14. Agricultural Reviews and Manuals, Southern Series No. 14.

Sonenshine, D. E. 1991. *Biology of ticks.* Vol. 1. New York: Oxford University Press.

Sonenshine, D. E. 1993. *Biology of ticks.* Vol. 2. New York: Oxford University Press.

Soulsby, E. J. S. 1965. *Textbook of veterinary clinical parastiology.* Philadelphia: Davis.

Soulsby, E. J. L. 1982. *Helminths, arthropods and protozoa of domesticated animals,* 7th ed. London: Balliere, Tindall & Cassell.

Steelman, C. D. 1976. Effects of external and internal arthropod parasites on domestic livestock production. *Ann. Rev. Entomol.* 21:155–178.

Stone, A., C. W. Sabrosky, W. W. Wirth, R. H. Foote, and J. R. Coulson. 1965. A catalog of the Diptera of America north of Mexico. Agriculture Handbook No. 276. U.S. Dept. Agriculture, Washington, DC.

Strickland, R. K., R. R. Gerrish, J. S. Hourrigan, and G. O. Schubert. 1976. Ticks of veterinary importance. Agriculture Handbook No. 485. Animal and Plant Health Inspection Service. Washington, DC: U.S. Department of Agriculture.

Strickman, D., J. E. Lloyd, and R. Kumar. 1984. Relocation of hosts by the sheep ked (Diptera: Hippoboscidae). *J. Econ. Entomol.* 77:437–9.

Sweatman, G. K. 1958. On the life-history and validity of the species in *Psoroptes*, a genus of mange mite. *Can. J. Zool.* 36:905–929.

Swingle, L. D. 1913. The life-history of the sheep-tick *Melophagus ovinus*. Agricultural Experiment Station Bull. 99. University of Wyoming, Laramie.

Tabachnick, W. J. 1996. *Culicoides variipennis* and bluetongue-virus epidemiology in the United States. *Ann. Rev. Entomol.* 41:23–43.

Theodorides, J. 1950. The parasitological, medical and veterinary importance of Coleoptera. *Acta. Trop.* 7:48–60.

Thomas, G. D., and S. R. Skoda, eds. 1993. Rural flies in the urban environment. North Central Regional Research Publication No. 335, Institute of Agriculture and Natural Resources, University of Nebraska, Lincoln.

Townsend, L., and P. Scharko. 1999. Lice infestation in beef cattle. *Comp. Count. Educ. Pract. Vet.* 21(Supplement):S119–123.

Tuff, D. W. 1977. A key to the lice of man and domestic animals. *Tex. J. Sci.* 28:145–159.

Uilenberg, G. 1994. Ectoparasites of animals and control methods. *Rev. Sci. Tech. Off. Int. Epizoot.* 13:979–1387.

U.S. Department of Agriculture. 1976. Ticks of veterinary importance. USDA-APHIS Agriculture Handbook No. 485. Washington, DC: U.S. Department of Agriculture.

U.S. Department of Agriculture. 1979. Proceedings of a workshop on livestock pest management. To assess national research and extension needs for integrated pest management of insects, ticks and mites affecting livestock and poultry. U.S. Dept. of Agriculture, EPA, and Kansas Agriculture Experiment Station.

U.S. Department of Agriculture. 1981. Status of biological control of filth flies. U.S. Dept. of Agriculture, SEA Publication A106.2:F64.

Van Bronswijk, J. E., and E. J. De Kreck. 1976. *Cheyletiella* (Acari: Cheyletiellidae) of dog, cat and domesticated rabbit, a review. *J. Med. Entomol.* 13:315–327.

Walker, A. R. 1994. *The arthropods of humans and domestic animals.* New York: Chapman & Hall.

Wall, R., and D. Shearer. 1997. *Veterinary entomology.* London: Chapman & Hall.

Wall, R., and D. Shearer. 2000. *Veterinary ectoparsites: biology, pathology and control,* 2nd ed. Abingdon, UK: Blackwell Science.

Walton, T. E., and B. I. Osburn. 1991. *Bluetongue, African horse sickness, and related arboviruses.* Boca Raton, FL: CRC Press.

Waterhouse, D. F. 1974. The biological control of dung. *Sci. Amer.* 230:100–109.

West, L. C. 1951. *The house fly. Its natural history, medical importance and control.* Ithaca, NY: Cornell.

West, L. S., and O. B. Peters. 1973. *An annotated bibliography of* Musca domestica *Linnaeus.* Folkstone, UK: Dawsons.

Wharton, R. H., and K. R. Norris. 1980. Control of parasitic arthropods. *Vet. Parasitol.* 6:135–164.

Whitaker, J. O., Jr. 1982. Ectoparasites of mammals of Indiana. Indiana Academy of Science, Monograph 4, Indianapolis: Indiana Academy of Science.

Whitworth, T. L. 2006. Keys to the genera and species of blow flies (Diptera: Calliphordae) of America north of Mexico. *Proceedings Entomological Society of Washington* 108:689–725.

Wikel, S. K. 1996. Host immunity to ticks. *Ann. Rev. Entomol.* 41:1–22.

Wilde, J. K. H., ed. 1978. *Tick-borne diseases and their vectors.* Tonebridge, Great Britain: Lewis Reprints.

Williams, J. R., and C. S. Williams. 1978. Psoroptic ear mites in dairy goats. *J. Amer. Vet. Med. Assoc.* 173:1582–1583.

Williams, J. F., and C. S. Williams. 1982. Demodicosis in dairy goats. *J. Amer. Vet. Med. Assoc.* 180:168–169.

Williams, R. E. 1985. Biological control of parasites. In S. M. Gaafar, W. E. Howard, and R. E. Marsh, eds., *Parasites, pests and predators,* pp. 87–102. Amsterdam: Elsevier Science.

Williams, R. E. 1986. Epidemiology and control of ectoparasites of swine. *Vet. Clin. North Am., Large Anim. Pract.* 2:469–480.

Williams, R. E. 1992. Control of flies in swine operations. *The Compendium.* May:689–692.

Williams, R. E., R. D. Hall, A. B. Broce, and P. J. Scholl, eds. 1985. *Livestock entomology.* New York: Wiley.

Wilson, B. H. 1968. Reduction of tabanid populations on cattle with sticky traps baited with dry ice. *J. Econ. Entomol.* 61:827–829.

INDEX

9 781138 118188